Refrigeration Systems and Applications

Refrigeration Systems and Applications

Special Issue Editors

Ciro Aprea
Angelo Maiorino
Adrián Mota Babiloni

MDPI • Basel • Beijing • Wuhan • Barcelona • Belgrade

MDPI

Special Issue Editors

Ciro Aprea
University of Salerno
Italy

Angelo Maiorino
University of Salerno
Italy

Adrián Mota Babiloni
Universitat Jaume I
Spain

Editorial Office
MDPI
St. Alban-Anlage 66
4052 Basel, Switzerland

This is a reprint of articles from the Special Issue published online in the open access journal *Energies* (ISSN 1996-1073) in 2019 (available at: https://www.mdpi.com/journal/energies/special_issues/ Refrig_Syst_Appl).

For citation purposes, cite each article independently as indicated on the article page online and as indicated below:

LastName, A.A.; LastName, B.B.; LastName, C.C. Article Title. *Journal Name* **Year**, *Article Number*, Page Range.

ISBN 978-3-03921-952-0 (Pbk)
ISBN 978-3-03921-953-7 (PDF)

Contents

About the Special Issue Editors

Ciro Aprea (Ph.D.) is a Full Professor of Applied Thermodynamics and the Research Supervisor of the Refrigeration Lab at the Department of Industrial Engineering of the University of Salerno (Italy). He oversees courses concerning energy and refrigeration technology for mechanical engineering. He is co-author of more than 90 international scientific papers. His research activities are focused on vapor compression systems, magnetic refrigeration, employment of carbon dioxide as a refrigerant, and the use of the phase change materials for cold storage.

Angelo Maiorino (Ph.D.) is an Associate Professor of Applied Thermodynamics and Senior Member of the Refrigeration Lab at the Department of Industrial Engineering of the University of Salerno (Italy). His courses give an overview of air conditioning and refrigeration plants for mechanical engineering and process unit operations for food engineering. He is co-author of more than 60 international scientific papers. His research activities are focused on vapor compression systems, magnetic refrigeration, employment of carbon dioxide as a refrigerant, and the use of phase change materials for cold storage.

Adrián Mota Babiloni (Ph.D.) is a Postdoctoral Researcher at the ISTENER Research Group of the Universitat Jaume I of Castellón (Spain). He is studying the adaptation of potential new low-global-warming refrigerants in refrigeration and air conditioning systems, organic Rankine cycles (ORC), and high-temperature heat pumps for waste heat recovery in an attempt to mitigate climate change.

Preface to "Refrigeration Systems and Applications"

Refrigeration applications are generally based on vapor compression systems and represent a significant contribution to global climate change. While refrigeration is instrumental to the development of humanity, it is predicted that an increase in the number of refrigeration applications will worsen the issue of climate change. Hence, energy-efficient systems with a lower contribution to global warming are required. In the last years the research and development of new working fluid technologies and methodologies have provided an opportunity for the transition from vapor compression systems based on fluorine fluids to more sustainable alternatives. For instance, the potential advantages and drawbacks of hydrofluoroolefins are being investigated, and mixtures with hydrofluorocarbons are being developed to find trade-off solutions.

Furthermore, the applications of hydrocarbons are being extended to installations that require a lower refrigerant charge. Lower flammability refrigerants require new flammability and risk analysis studies to determine their possible hazard. Heat and mass transfer phenomena studies are being carried out for new pure and mixed refrigerants. Ejectors are being studied to increase energy performance in particular applications. Alternative technologies based on renewable energy or solid states, such as solar cooling or magnetic refrigeration, are being developed and integrated into new processes. The integration of phase change materials and slurries is a promising new alternative. Finally, nanoparticles and nanofluids have opened an entirely new world of possibilities.

The available literature on these topics is still in its early stages and these working fluids, technologies, and methodologies are not considered mature. However, there is significant potential to improve energy efficiency as well as the operation and capacity of these new approaches.

Ciro Aprea, Angelo Maiorino, Adrián Mota Babiloni
Special Issue Editors

energies

MDPI

Article

Thermal and Energy Evaluation of a Domestic Refrigerator under the Influence of the Thermal Load

Juan M. Belman-Flores [1,*], Diana Pardo-Cely [1], Miguel A. Gómez-Martínez [1], Iván Hernández-Pérez [2], David A. Rodríguez-Valderrama [1] and Yonathan Heredia-Aricapa [1]

[1] Engineering Division, Campus Irapuato-Salamanca, University of Guanajuato, C.P. 36885 Salamanca, Mexico; dianapardocely@gmail.com (D.P.-C.); gomezma@ugto.mx (M.A.G.-M.); davidalejandrorv@gmail.com (D.A.R.-V.); yonheredia@hotmail.com (Y.H.-A.)
[2] División Académica de Ingeniería y Arquitectura, Universidad Juárez Autónoma de Tabasco, C.P. 86690 Cunduacán, Mexico; ivan.hernandezp@ujat.mx
* Correspondence: jfbelman@ugto.mx; Tel.: +52-464-647-9940

Received: 4 December 2018; Accepted: 22 January 2019; Published: 27 January 2019

Abstract: This study seeks to understand the thermal and energetic behavior of a domestic refrigerator more widely by experimentally evaluating the main effects of the thermal load (food) and the variation of the ambient temperature. To carry out the experiments, the thermal load was classified based on the results of a survey conducted on different consumers in the state of Guanajuato, Mexico. The thermal behavior of both compartments of the refrigerator, the total energy consumption, the power of the compressor in its first on-state, and the coefficient of performance, according to the classification of the thermal loads and the room temperature, were evaluated. Finally, it is verified that the thermal load and the room temperature have a significant influence on the energy performance of the refrigerator.

Keywords: energy consumption; thermal load; domestic refrigeration system

---

1. Introduction

The domestic refrigerator is one of the most popular household appliances because of its use in food preservation. Most of these refrigerators are based on vapor compression technology, and their continuous operation represents a high-energy consumption. Currently, it is claimed that the refrigeration sector (including air conditioning) consumes about 17% of the total electricity used worldwide, where there are currently more than 1.5 billion domestic refrigerators in use [1].

In Mexico, approximately 86% of households have at least one refrigerator, representing more than 28 million domestic refrigerators in use [2]. According to the Trust for Saving Electrical Energy (FIDE, from its Spanish initials), the refrigerator represents around 30% of the total energy consumption in a household [3]. For several decades, there was an imminent growth in the refrigeration industry, which also led to a considerable increase in energy consumption. Thus, these appliances are a point of interest in search of energy improvements. Some methods, such as energy labeling, take into account the efficiency of the product [4], which guarantees to some extent the regulations on energy saving. Thus, the labeling provides a guide to study different mechanisms that can increase a refrigerator's energy efficiency, such as the design of the main components, thermal insulation, adequate thermal behavior, and use of alternative refrigerants, among others [5]. However, the refrigerators' energy consumption does not only depend on the technical characteristics of the components, but it also depends on the usage habits of the consumer and the environmental conditions where the appliance is located, as it is specified on the energy label [6].

There are factors such as ambient or room temperature [7], relative humidity [8], and frost formation, [9], among others, that significantly affect the energy performance of a domestic refrigerator. In addition to the above, other factors depend on the usage habits of the consumer [10], who plays a significant role in the cold chain and the proper conservation of food. Among these factors, the following can be mentioned: the frequency in the opening of doors, the position of the thermostat, the amount of food, and the cleaning in the case of external condensers, among others.

In the literature, there are works on the study of these factors; for instance, Saidur et al. [11] experimentally evaluated the temperature of the room and other factors, such as the opening of doors and the position of the thermostat, on the energy consumption of a refrigerator. The authors concluded that the temperature of the room affects, to a greater extent, the energy consumption, followed by the opening of doors. Hasanuzzaman et al. [12] analyzed the energy consumption of a domestic refrigerator by varying factors such as the number of door openings, opening duration, cabinet load, thermostat position, and room temperature. The authors found that all factors influence energy consumption, with the most notorious case (with a 40% increase) when the refrigerator operates with open doors compared to when it is used with closed doors. Later, the authors extended their study to analyze factors such as the position of the thermostat, the thermal load, and the ambient temperature on the heat transfer and the energy consumption of the refrigerator. The authors concluded that the largest contributions occurred when the thermal load varied from 0 to 12 kg, with an increase in energy consumption of 58%, and when the ambient temperature changed from 18 °C to 30 °C, with a 41% increase [13]. Khan et al. [14] presented another work, similar to the previous studies, confirming an increase in energy consumption of up to 30% depending on the frequency of door opening, an increase of 30% when the ambient temperature varied from 20 °C to 30 °C, and an increase of 59% when the load varied from 0 m^3 to 0.007 m^3.

In the literature, there are also works with a statistical approach based on a series of surveys and related to the usage habits of consumers. For instance, Janjic et al. [15] investigated the conditions, such as temperature, cleanliness, and storage practices, under which food is subjected inside refrigerators. The authors reported that about half of the refrigerators considered in the survey had an incorrect food storage practice. Furthermore, the internal temperature of the refrigerator was considered to be high compared to the recommended temperature for this household appliance. Geppert and Stammiger [16] evaluated the behavior of the consumer in relation to the use of the refrigerator and the main characteristics of these appliances. They analyzed the conditions of the ambient temperature, the internal temperature of the compartment, and the heat sources near the refrigerator, aspects that influence the thermal and energetic performance of the household appliance. Based on the results, the authors made a series of recommendations on energy efficiency, and concluded that there is a lack of information provided to the consumers on this subject. Later, the authors extended their study to experimentally evaluate some of the operational factors that reflect the daily use of refrigerators such as the ambient temperature, the position of the thermostat, and the thermal load influenced by the amount of food. They concluded that the energy consumption is very sensitive to the ambient temperature and, to a lesser extent, the internal temperature of the refrigerator and the thermal load [17]. On the other hand, James et al. [18] made a review of diverse works where they analyzed factors such as the frequency in the opening of doors, the cleaning, the handling and storage of food, and the age of the refrigerator. This compilation was carried out aiming to analyze the thermal behavior of the refrigerator and the cleaning of the food on the impact on the consumers' health. Thus, it is clear that factors, where the consumer is involved, reflect in a meaningful way the energetic and thermal behavior of the refrigerator, together with the environmental conditions where the appliance is located.

The literature review indicates the importance of external factors on the energy consumption of domestic refrigerators, and it also shows that those factors are unrelated to the design of the refrigerator components. One of the factors that affects the energy consumption to a great extent is the room temperature where the appliance is located. Other factors of importance in the energetic operation of the refrigerator are also those related to the usage habits of the consumer. On the other hand, in the studies found in the literature, there is no justification for the thermal load (food) evaluated. In addition, the results presented with variation in thermal load focus only on the energy performance of the refrigerator.

In this paper, with knowledge based on surveys on usage habits, the energy consumption and the average temperature in both compartments of a domestic refrigerator are evaluated when the thermal load (food) is varied. Moreover, the effect of room temperature on refrigerator performance is analyzed, with both factors (thermal load and temperature) recorded in the surveys. Thus, this paper provides a basis for a deeper analysis and a better understanding of the energy consumption of a refrigerator. This type of study should facilitate recommendations through the manufacturer, from an energy and thermal viewpoint, on how to better use the appliance based on the amount of food and, in general, on the habits of use that cause great increases in energy consumption and which can degrade food quality due to inappropriate temperatures in the compartments. As an additional contribution, this study provides information to consumers and manufacturers as a reinforcement to understand how refrigerator usage habits affect thermal conditions and energy consumption and, thus, improve the refrigerator use recommendations.

The rest of the paper is organized as follows: in Section 2, the thermal load obtained from surveys is shown. In Section 3, the experimental refrigerator and the tests performed are presented. Section 4 shows the main results of thermal behavior of compartments and energy consumption of the refrigerator. Finally, Section 5 summarizes the main conclusion of the study.

2. Presence of Thermal Load

Among the different factors influencing the proper performance of a domestic refrigerator, the adequate distribution of airflow in the compartments is highlighted, which has an impact on thermal behavior and, in turn, affects energy consumption. In this respect, the thermal load (food stored in the refrigerator's compartments) also influences the thermal behavior, where the consumer plays a key role in the use of this appliance.

The thermal loads experimentally evaluated in this work are based on the records of surveys applied to 200 random consumers in Salamanca, Guanajuato, Mexico. Along with the questions asked to analyze the use of the refrigerator regarding the thermal load, and with previous consent of the respondents, visual evidence was collected, as well as the measurement of the amount of food stored in the fresh-food compartments (crisper drawers) and in the freezer. Figure 1 shows the conditions of thermal load in both compartments, for which the filling of the refrigerator was classified in four ranges, as shown in the figure. The light-blue color corresponds to the thermal load in the food compartment, whereas the dark-blue color represents the amount of food in the freezer.

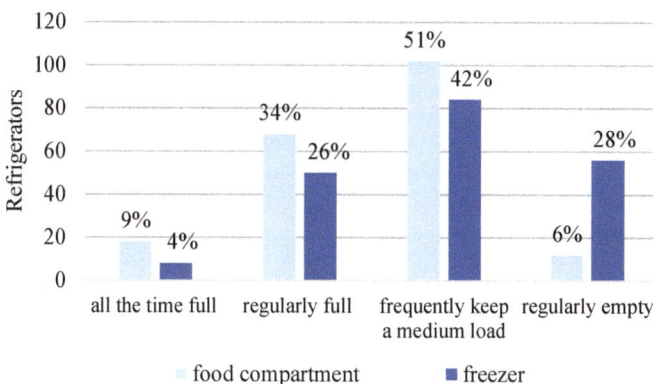

Figure 1. Distribution of thermal load in the refrigerator.

The higher percentages correspond to the consumers frequently keeping their refrigerator at a medium load—51% of consumers (102 refrigerators) for the food compartment and 42% (84 refrigerators) for the freezer. On the other hand, a low percentage of consumers keep their refrigerator full all the time—9% for the food compartment and 4% for the freezer. According to the statistic shown in Figure 1, an average of eight thermal loads were defined in this work (see Table 2). Moreover, during the surveys, the temperature of the room where the refrigerator was located was also measured; this way, an average temperature sample was set during some experimental tests to analyze their effect on the refrigerator's performance.

3. Experimental Refrigerator

According to the surveys, it was observed that a great percentage of consumers have medium-sized refrigerators at home (two-doors and no-frost type). For this reason, a refrigerator meeting most of the features of the survey's refrigerators was used, as shown in Figure 2. The two-door experimental refrigerator had a volume capacity of 0.3 m³ (300 L), separating the fresh-food compartment at the bottom and the frozen-food compartment at the top. The refrigerator was a no-frost type and the heat transfer in the freezer occurred via forced convection. Table 1 shows more general features of the experimental refrigerator.

Table 1. General features of the experimental refrigerator.

External Dimensions	
Width	0.54 m
Length	0.67 m
Height	1.64 m
Net weight	52.4 kg
System Characteristics	
Refrigeration	Forced convection
Melting element	By electrical resistance
Defrost	Automatic
Refrigerant	R134a
Voltage/Current	127~/60 Hz/1.1 A

3.1. Instrumentation and Measurements

The refrigerator was used in this research to evaluate the thermal behavior of the compartments, as well as the energy consumption when the thermal load varied in both compartments according to the surveys. To measure the temperature, 15 J-type thermocouples were used with an uncertainty of measurement of ±0.3 K. Eleven thermocouples were distributed in the food compartment and were located within containers of 0.245 L with a mixture of 50% water and 50% glycol. Four thermocouples were placed in the freezer inside wooden cubes, due to their high capacity of humidity absorption, thus allowing a constant measurement of the temperature. In Figure 2a, the distribution of the thermocouples in both compartments is illustrated; moreover, the distribution of the water compartments can be seen, simulating the thermal load for a specific case. On the other hand, to measure the energy consumption, a Fluke 1735 energy logger (Fluke, Everett, WA, USA) calibrated with a measurement error of ±1.5% was utilized.

The thermocouples were connected to an NI-9213 card attached to the chassis NI cRIO-9030 (National Instruments (NI), Austin, TX, USA). Via a Universal Serial Bus (USB) connection to a computer, a real-time visualization was possible with the SignalExpress software (National Instruments (NI), Austin, TX, USA) programmed in LabView. The temperature measurement was recorded in intervals of 10 seconds, whereas the measurement of the energy consumption was set in intervals of one minute; the data were stored on a Secure Digital (SD) card. Both the temperature and energy consumption measurements were done simultaneously.

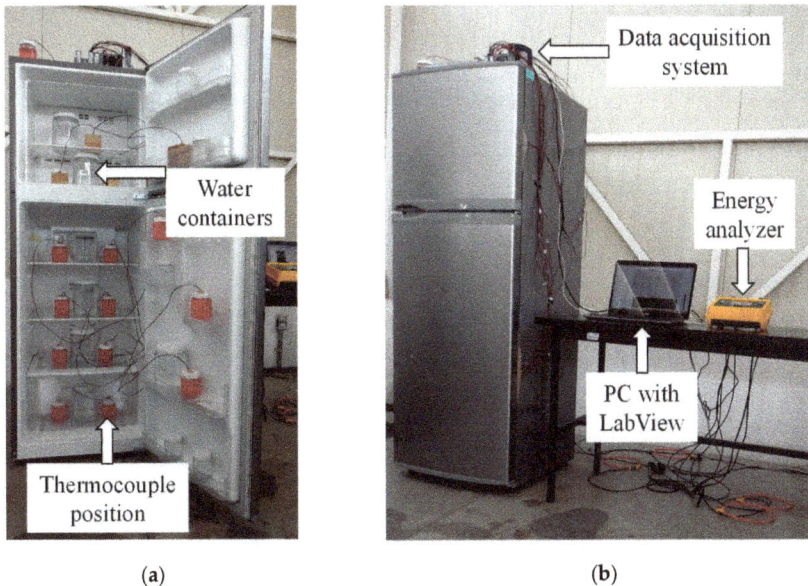

(**a**)   (**b**)

Figure 2. Experimental test bench: (**a**) temperature distribution; (**b**) instrumentation.

3.2. Proposed Tests

As mentioned before, the aim of this study was to evaluate the effect of the thermal load (food) on the thermal and energy behavior of a domestic refrigerator. In this sense, the foods were simulated with containers full of water and whose volume capacities were 0.3, 1, 1.8, and 4 L.

According to the information gathered in the surveys, different ranges of thermal load were classified (see Figure 1), where the total variation of the average thermal load (food compartment and freezer) went from a minimal load of 7 kg (Regularly empty, 5 kg in the fresh-food compartment and 2 kg in the freezer) to a maximum load of 39 kg (All the time full, 27 kg in the fresh-food compartment and 12 kg in the freezer). Additionally, ambient temperatures of 20 °C and 25 °C, with a variation in intervals of ±0.5 °C, were frequently measured in the room (giving to surveys) where the refrigerator was located and, in relation to these ambient temperatures, the loads were also grouped. The above data can be observed in Table 2, where a reference condition is included, that is, when the refrigerator remains empty.

Table 2. Thermal loads in both compartments under two conditions of room temperature.

Room Temperature	20 °C		25 °C	
Thermal Load	**Fresh-Food Compartment (kg)**	**Freezer (kg)**	**Fresh-Food Compartment (kg)**	**Freezer (kg)**
Reference	0	0	0	0
Regularly empty	5	2	11	2
Frequently keep at medium load	27	2	18	2
Regularly full	27	5	18	7
All the time full	27	7	27	12

All the tests were performed in the same way. Firstly, the refrigerator was loaded with a certain amount of food, as shown in Table 2. Once the refrigerator was loaded, the test initiated with the start-up of the refrigerator and at the corresponding room temperature of the load, according to Table 2. Note that, for each test, the refrigerator and the thermal load were at room temperature. The test continued until the thermal stability was reached in both compartments and, during the test, the doors of the refrigerator were kept closed. Also, the damper (control element) remained in the fifth position, exactly as it was when the refrigerator left the factory. After finishing the test, the refrigerator was unplugged and defrosted so that the refrigerator could reach room temperature.

4. Results and Discussion

In this section, the main results coming from the thermal behavior of both compartments of the refrigerator are presented, as well as the energy consumption for different conditions of the thermal load and room temperature. Each test was done in triplicate, aiming to yield greater reliability in the results, which reflect the average of the temperature and energy measurements. Moreover, the presented results are those obtained when the thermal stability was achieved in both compartments.

4.1. Effect of Thermal Load on Thermal Behavior of the Compartments

Figure 3 shows the conditions of temperature in both compartments of the refrigerator for a room temperature of 20 ± 1 °C. The compartment temperature represents the average of the thermocouples placed within them. The horizontal axis of the figure represents the thermal loads, where 0 kg corresponds to an empty refrigerator (without thermal load) in both compartments, and 7, 29, 32, and 34 kg correspond statistically to the average load (fresh-food compartment and freezer) of each of the classifications of the thermal load shown in Figure 1 and Table 2. The light-blue color represents the temperature of the food compartment, and the dark-blue color represents the temperature of the freezer. In Figure 3, it is observed that the temperature of the food compartment showed relatively small changes as the thermal load increased, remaining at a maximum difference of 2 °C between loads 7 and 29 kg. Furthermore, the freezer experienced a maximum thermal variability of 4 °C between the loads of 7 and 29 kg.

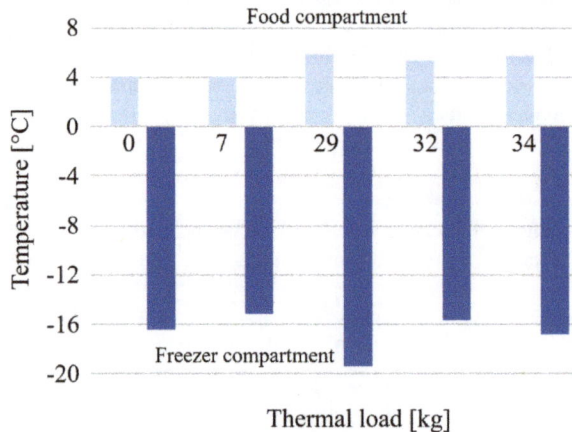

Figure 3. Thermal behavior of the compartments at 20 °C.

On the other hand, in Figure 4, the thermal behavior of both compartments at different thermal loads and at a room temperature of 25 ± 1 °C is illustrated. It is worth mentioning that these thermal loads are the most representative for the room temperature measured in the surveys. It can be observed in the figure that both compartments represented a variable thermal condition, without having a clear correspondence between the thermal load and the compartment temperature. The maximum temperature variation in the food compartment was 2.2 °C (0 and 20 kg), while, in the freezer, it was 2.7 °C (0 and 25 kg).

Consistent with these behaviors, it can be confirmed that the refrigerator is capable enough to maintain an operational range of adequate temperatures in both compartments, regardless of the amount of thermal load.

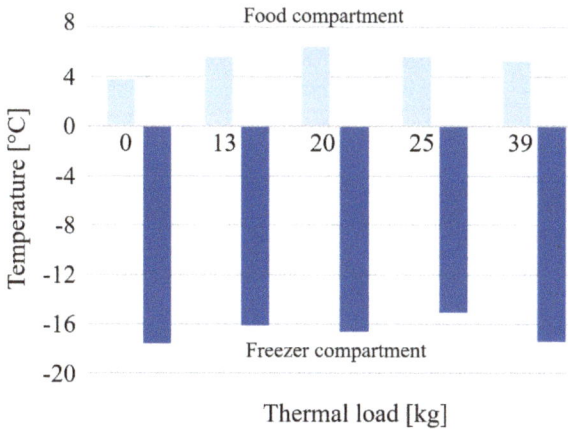

Figure 4. Thermal behavior of the compartments at 25 °C.

4.2. Effect of Thermal Load on Energy Consumption

The cooling capacity of a refrigerator is directly proportional to the cabinet inner thermal load (mass), which depends on the food initial temperature, the cabinet temperature, the specific heat, and the latent heat of the thermal load (water). Moreover, this mass is heated during the off-state of the compressor for cooling again during the on-state. Therefore, the energy consumption must increase as the thermal load in the refrigerator increases. In this sense, Figure 5 shows the total energy consumed by the refrigerator for the different thermal loads. Moreover, the energy behavior is shown for two conditions of room temperature. It can be clearly observed that, when the thermal load increased, the energy consumption also increased; these are similar behaviors found by References [12,14]. In Figure 5, it can be noticed that the magnitude of energy behavior at a temperature of 25 °C was higher than at a room temperature of 20 °C. Here, it is clear that the increase in room temperature caused an increase in the thermal leap between the ambient and the cabinet; thus, a significant amount of heat was transferred via conduction through the refrigerator's walls. For example, in Figure 5 it can be observed that, for the reference load (0 kg), there was an increase of 0.4 kWh for a temperature condition ranging from 20 °C to 25 °C; on the other hand, for the ambient condition of 20 °C, there was an increase ranging from 0.4 kWh (0 kg) to 3.5 kWh (34 kg); this values ranged from 0.8 kWh (0 kg) to 4.5 kWh (39 kg) for the temperature of 25 °C. Note that these energy consumptions vary in accordance with the thermal stabilization time of each test (see Table 3). Based on Figure 5, it was concluded that the thermal load represents a strong influence on the refrigerator's energy consumption. Finally, it can be said that, as the thermal load increases, so does the evaporation temperature. Therefore, the refrigeration cycle responds according to the evaporation temperature.

In Table 3, more information about the refrigerator's energy behavior is provided. It can be noted that, for the reference test (0 kg), the time estimated to reach thermal stability increased around 4 h for a condition of ambient temperature (room temperature) fluctuating from 20 °C to 25 °C. This increase caused the switch-on (on-state) percentage of the compressor to rise to 4%, which represents an increase of 0.029 kWh per operating hour. On/off cycles of the compressor clearly evidence the thermal behavior of the refrigerator compartments, which is linked to the temperature control in relation to the position of the damper. It is, therefore, consistent that the time of thermal stability increases as the thermal load increases and due to the increase in ambient temperature. With regard the work cycles shown in the table, a correlation referring to the load increase does not exist. Note that the percentage switch-on and the cycles decreased as the thermal load increased (e.g., from 29 to 32 kg (20 °C) and from 20 to 25 kg (25 °C)). For these conditions, the thermal load of the food compartment remained constant, while the freezer load increased in each test (see Table 2). Note that, for this particular refrigerator, the compressor regulation work is linked to the temperature of the food compartment and to the temperature of the freezer.

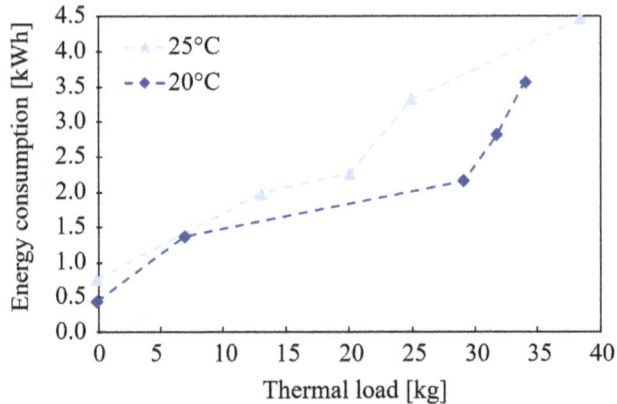

Figure 5. Energy consumption for different thermal loads.

Table 3. Energy behavior at different loads and constant ambient temperature.

Room Temperature (°C)	Thermal Load (kg)	Thermal Stability Time (h)	% Switch-On	Total Energy (kWh)	Cycles (24 h)
	0	8	32	0.4	24
	7	21	35	1.4	21
20	29	24	40	2.2	27
	32	33	37	2.8	24
	34	38	38	3.6	25
	0	12	36	0.8	24
	13	25	39	2.0	34
25	20	25	42	2.3	34
	25	38	37	3.3	26
	39	46	36	4.5	24

4.3. Effect of Thermal Load on the First On-State of the Compressor

The stage consuming the most energy in a household refrigerator originated when the foods were stored for their cooling. Therefore, it is recommended that this process be quick to avoid inappropriate conservation. For this reason, the first on-state of the compressor when the refrigerator is started is larger than the following ones. This occurs when the refrigerator contains too much thermal load (food), as shown in Figure 6. The figure shows the power of the compressor for the different thermal loads mentioned above. Figure 6a corresponds to a room temperature of 20 °C, and Figure 6b corresponds to a temperature of 25 °C. In both figures, it is clearly evident that the power input of the first on-state was linked to the amount of food stored in the refrigerator. Some studies mentioned that the additional energy consumption originated during the food-cooling stage [17]. This cooling stage is particularly evident during the first on-state of the compressor. In both figures, at the beginning of each cycle, there was a high-power peak due to the normal behavior of the electric motor. In addition, an increase in power was observed as the load increased; this conditions the on/off cycles of the compressor, requiring greater power to lower the temperature for a greater quantity of food.

(a)

(b)

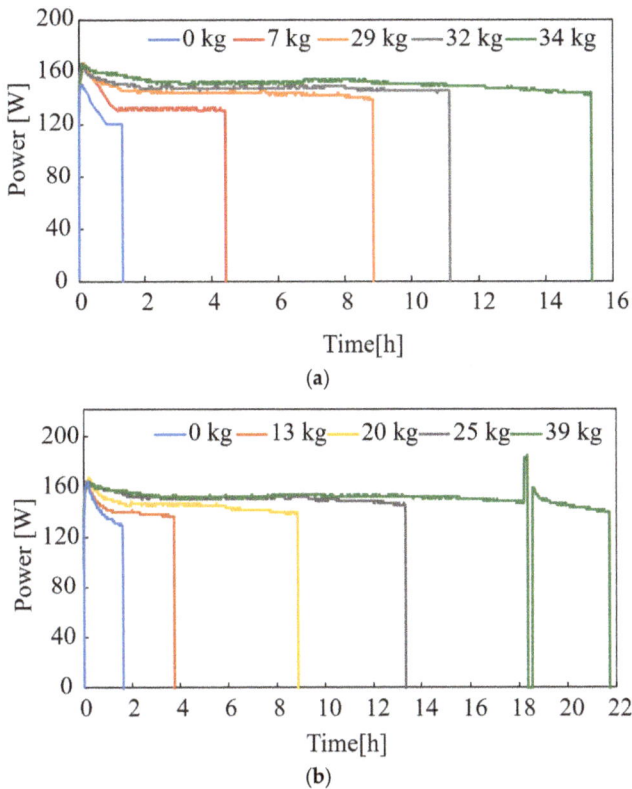

Figure 6. Power during the first on-state of the compressor: (**a**) room temperature at 20 °C; (**b**) room temperature at 25 °C.

As mentioned before, the ambient temperature affects the refrigerator's energy behavior to a large extent, as can be seen in Figure 6a,b. For the specific case of the reference load (0 kg), there was a difference in the average of the power of approximately 11 W, which indicates a power increase when the room or ambient temperature increased by 5 °C. This reflects an increase of 0.05 kWh in energy consumption. Thus, it is well known that the domestic refrigerator's electricity consumption is very sensitive to the ambient temperature [11]. Another aspect to observe in Figure 6b is that the thermal load of 39 kg represented the condition that consumed the most energy and whose on-time of the compressor was approximately 21 h. Moreover, in this case, it can be noticed that, at around 18 hours, a defrost occurred, indicated by the power increase, which in turn caused the compressor to shut down. Finally, it can be concluded that the on-time of the compressor on the first start increased when the thermal load increased; this behavior was reflected in the total energy consumption of the refrigerator.

4.4. Effect of Ambient Temperature on Thermal Behavior of the Compartments with a Constant Thermal Load

In order to expand this study based on survey data, in Figure 7, the average temperature of the food compartment for a constant thermal load and under two different conditions of room temperature is illustrated. The fading of the gray color in the bars represents a condition of low room temperature, whereas the discoloration of the blue color represents a higher room temperature. In Figure 7, it can be observed that, for a certain load, an increase in the room temperature caused a rise in the temperature of the fresh-food compartment (FF). For example, for the load of 13 kg, the increase from 16 to 25 °C in room temperature caused an increase of approximately 1.5 °C in the food compartment. Note that the foregoing is based on the thermal stability in the food compartment and for a fixed condition of the damper set at 5. For the thermal loads of 20 and 23 kg, a very similar thermal behavior existed, where the average temperature of the compartment experienced an increase for both room temperatures, showing an average thermal difference of 0.6 °C in both cases. Finally, for the thermal load of 30 kg, it can be noted that the average temperature reached in the food compartment for both room conditions was almost the same, around 6 °C.

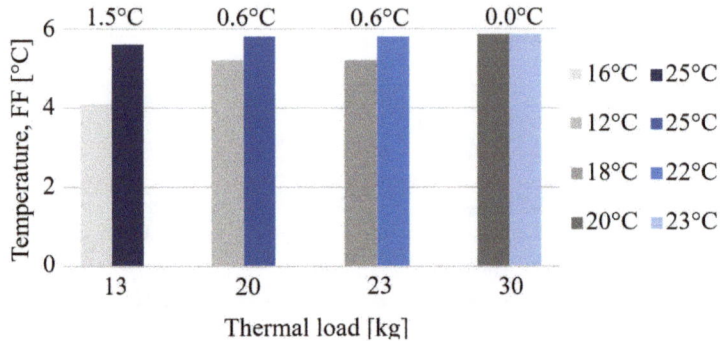

Figure 7. Temperature in the fresh-food compartment maintaining the constant thermal load at two ambient temperature conditions.

With respect to the average thermal behavior of the freezer (FZ), Figure 8 shows the ambient temperature effect for constant thermal loads. It can be seen that, when a constant thermal load was subjected to two room temperatures, the lowest average freezer temperature corresponded to the highest room temperature condition. For example, for the loads of 13, 20, and 23 kg, a difference in the temperature of 2, 2.3, and 2.6 °C, respectively, was obtained in this compartment, observing the effect of the room temperature rise. For the load of 30 kg, as also seen in Figure 7, the thermal difference was practically very small (about 0.6 °C).

Based on the reported results in this section and in the previous sections, it can be said that, for different thermal loads and different ambient temperatures, the refrigerator has the capacity to maintain an adequate thermal behavior for food preservation. However, this is achieved with a variation in energy consumption according to the ambient condition and the habit of use in the thermal load management.

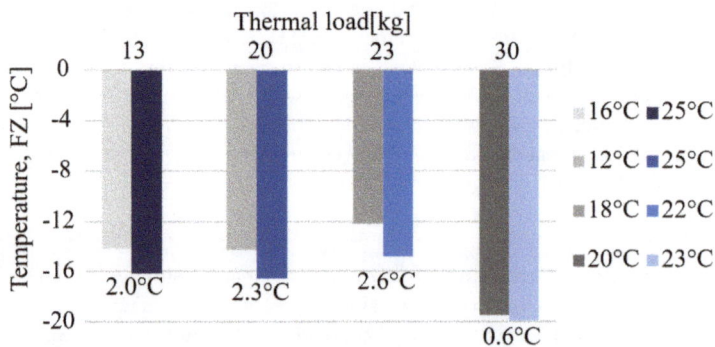

Figure 8. Temperature in the freezer compartment maintaining a constant thermal load at two temperature conditions.

4.5. Effect of Ambient Temperature on Energy Consumption with a Constant Thermal Load

Figure 9 shows the energy consumption when the refrigerator worked with constant thermal loads and at two room temperature conditions. The fading in gray and blue colors indicates the room temperatures. The energy increase percentage for each thermal load subjected to two conditions of ambient temperature is indicated at the top of the bars. It can be clearly observed that the increase of the ambient temperature caused an increase in energy consumption. For example, the load of 20 kg represented a greater increase in energy consumption, around 72% for an ambient thermal difference of 13 °C (room temperature between 12 °C and 25 °C). For this increase in the ambient temperature, the energy consumption of the refrigerator varied from 1.3 kWh/day to 2.3 kWh/day, a rise of 73 Wh/day per °C.

In the literature, there are works that reported information about the energy consumption for refrigerators of volumetric capacity different from the one evaluated in this work. The range of thermal load was also different. In order to show the results and compare the orders of magnitude, Hasanuzzaman et al. [12] experimentally concluded that, when the ambient temperature experienced an increase of 12 °C, the energy consumption varied from 2.13 kW/day to 3.64 kW/day, increasing around 126 Wh/day per °C rise in the ambient temperature of a 460-L refrigerator. Masjuki et al. [8] found that, when the ambient temperature increased 15 °C (from 16 to 31 °C), the energy consumption increased from 0.56 kWh/day to 1.12 kWh/day, representing an energy consumption around 40 Wh/day per °C rise in the ambient temperature for a 150-L refrigerator. Meier [19] concluded that in an increase in the ambient temperature of 11 °C (from 17 °C to 28 °C), the energy consumption varied from 1.25 kWh to 2.6 kWh/day. Finally, based on Figure 9, it was concluded that there is a strong influence of the ambient or room temperature on the domestic refrigerator's energy consumption.

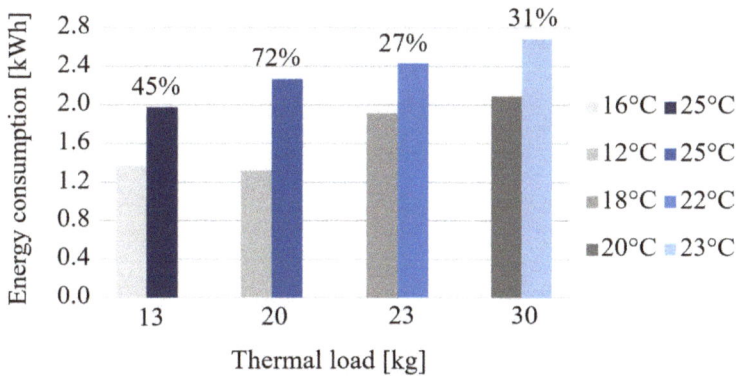

Figure 9. Energy consumption at different conditions of ambient temperature.

On the other hand, the quantity of heat to be removed can be estimated from knowledge of the thermal load (water), including its initial state at the entrance of the compartments, final state, mass, specific heat above and below freezing temperature, and latent heat [8]. Thus, the coefficient of performance (*COP*) can be calculated as follows for heat removed from the food compartment:

$$Q_{FF} = m_{FF} C p_{water} (T_{amb} - T_{FF}). \tag{1}$$

That for heat removed from the freezer can be calculated as follows:

$$Q_{FZ} = Q_{FZ_1} + Q_{FZ_2} + Q_{FZ_3}; \tag{2}$$

$$Q_{FZ_1} = m_{FZ} C p_{water} (T_{amb} - 0°C); \tag{3}$$

$$Q_{FZ_2} = m_{FZ}h_f;$$ (4)

$$Q_{FZ_3} = m_{FZ}Cp_{frost}(0°C - T_{FZ}).$$ (5)

The capacity of the refrigeration system for the thermal load can be determined from the time set aside for heat removal:

$$CC = \frac{Q_{FF} + Q_{FZ}}{3600\ t}.$$ (6)

Therefore, the *COP* can be estimated as

$$COP = \frac{CC}{E/t}.$$ (7)

Figure 10 illustrates the behavior of the coefficient of performance (*COP*) of the domestic refrigerator under study. It can be seen that, for a constant load, there was a decrease in *COP* as the room temperature increased. For example, for a thermal load of 13 kg, there was a decrease in *COP* of 10% when going from 16 °C to 25 °C. An increase in room temperature reduced the overall refrigeration system *COP* by increasing the difference between the evaporating and condensing temperature.

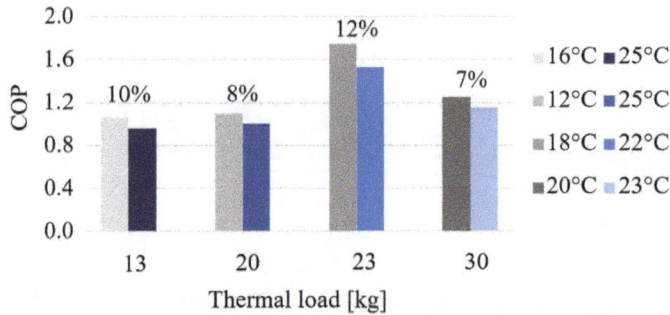

Figure 10. Coefficient of performance at different conditions of ambient temperature.

Table 4 shows the energy behavior for the evaluated thermal loads in this section. It can be noted that an increase in ambient or room temperature also caused a rise in the thermal stabilization time and, thus, the percentage of the compressor's start-up increased, resulting in an increase in the total energy consumption. In other words, to maintain the desired temperature, the on/off frequency of the compressor (cycles), as well as the starting time, increased. This happened because the energy consumption increased with the rise in room temperature. Another existing behavior was that when the thermal load rose, and under ambient temperature conditions lower than 20 °C, there was an increase in the number of cycles in one day; this means that the on/off cycles of the compressor were shorter and more common. This is because the change rate of the refrigerator temperature is linked to the food thermal capacity. When the refrigerator is full of food, it causes a greater loss of energy due to the on/off cycles. Therefore, energy consumption increases when thermal load increases, to be able to control the desired temperature.

In Figure 11, the power of the first start of the compressor is illustrated, for the load variation shown in Table 4. Figure 11a shows the difference in both the starting time and the power for a thermal load of 13 kg, which presented an increase from 16 °C to 25 °C in the ambient temperature. This behavior was also observed in the other thermal loads subjected to different conditions of room temperature. For example, for the thermal loads of 20 kg and 30 kg (Figure 11b,d), a greatest increase in the compressor operation time, 5 h approximately, was presented. The thermal load of 30 kg appears to be the condition that prolonged this time the most for the two conditions of room temperature, representing an increase of 0.62 kWh. Hence, the influence of the ambient temperature on the consumed power by the compressor on the first start was confirmed with these behaviors.

Table 4. Energy behavior for different thermal loads.

Thermal Load (kg)	Room Temperature (°C)	Thermal Stability Time (h)	% Switch-On	Total Energy (kWh)	Cycles (24 h)
13	16	24	30	1.4	19
	25	25	39	2.0	34
20	12	24	27	1.3	23
	25	25	42	2.3	34
23	18	22	33	1.9	28
	22	26	35	2.4	33
30	20	24	40	2.1	27
	23	25	44	2.6	23

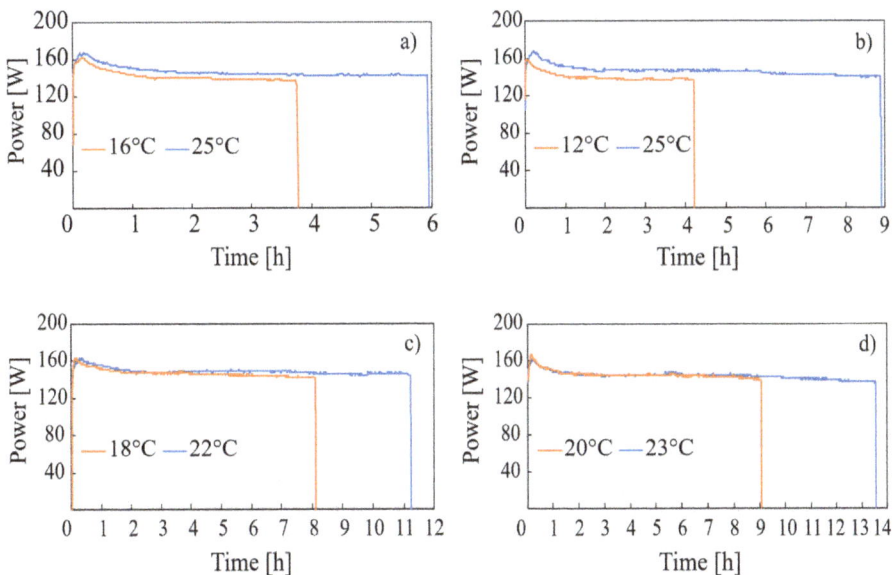

Figure 11. Consumed power in the start of the compressor for different thermal loads: (a) 13 kg; (b) 20 kg; (c) 23 kg; (d) 30 kg.

In accordance with the previous sections, the existing importance of the amount of food with regards to the refrigerator energy consumption was confirmed and, to a lesser extent, the effect that this causes in the compartment's thermal behavior. This type of refrigerator works within a range of adequate temperatures for food preservation.

According to the surveys done before this work, only 20% of consumers purchase a refrigerator with the ideal storage capacity (food storage capacity). On the other hand, 100% of consumers ignore the effect of the amount of food and ambient temperature on the appliance performance. In this sense, this type of study should facilitate recommendations through the manufacturer from an energy and thermal viewpoint on how to better use the appliance based on the amount of food and, in general, on the habits of use that cause great increases in energy consumption and that can degrade the food quality due to inappropriate temperatures in the compartments.

Finally, based on the experimental data, a multiple linear regression equation was developed, which shows the combined effect of the ambient temperature (T_{amb}) and the thermal load (L) on the energy consumption (E) for this refrigerator. This equation has a correlation degree of 85%.

$$E = -1.55 + 0.0395m + 0.104T_{amb}. \tag{8}$$

5. Conclusions

In this work, the effect of the thermal load (food) on energy consumption and the thermal condition of the compartments of a domestic refrigerator was experimentally evaluated. The amount of the evaluated thermal load was based on previous surveys applied to consumers to define the amount of food in both compartments. The main objective of this work was to provide more information about domestic refrigeration in term of the effect of the thermal load and the room temperature on the refrigerator performance. For this, a typical refrigerator of 0.3 m^3 was used, where the tests were conducted in triplicate for each assay to achieve reliability. Among the most relevant aspects in this study, the following are mentioned:

- A survey was applied to determine the amount of food commonly stored in the refrigerator compartments. This way, the different thermal loads (from a minimal load of 7 kg to a maximum of 39 kg) were evaluated in this work, an aspect which was not previously justified in the literature.
- The thermal behavior in both compartments of the refrigerator did not show any correspondence with respect to the thermal load increase, while keeping a constant room temperature.
- It was observed that the energy consumption had a great effect due to the thermal load increase where, for example, for an ambient condition of 20 °C, an increase from 0.4 kWh (0 kg as reference) to 3.5 kWh (34 kg) was observed; this increase was from 0.8 kWh (0 kg) to 4.5 kWh (39 kg) for a temperature of 25 °C.
- The first on-state of the compressor increased when the thermal load increased; this behavior was reflected in the refrigerator's total energy consumption.
- For a constant thermal load, the increase in room temperature caused an increase in the temperature of the fresh-food compartment. For a load of 13 kg, the increase in room temperature from 16 to 25 °C caused a rise of 1.5 °C in the food compartment's average temperature. For the case of the freezer, a maximum decrease of 2.6 °C was obtained for a load of 23 kg, whose room temperature increased from 18 to 22 °C.
- Finally, increases in the refrigerator's energy consumption were clearly observed for a constant thermal load and an increase in room temperature.

Author Contributions: Conceptualization, J.M.B.-F. and D.P.-C.; methodology, D.A.R.-V. and Y.H.-A.; resources, M.A.G.-M.; analysis of data, I.H.-P., J.M.B.-F. and D.P.-C. wrote the paper. All authors have read and approved the final manuscript.

Funding: This research received no external funding.

Acknowledgments: We acknowledge the University of Guanajuato for their sponsorship in the realization of this work.

Conflicts of Interest: The authors declare no conflict of interest.

Nomenclature

CC	Cooling load (kW)
COP	Coefficient of performance
Cp	Specific heat (kJ/kgK)
E	Energy consumption (kWh)
h_f	Latent heat of fusion of the of the water (kJ/kg)
m	Thermal load (kg)
Q	Heat removal (kJ)
T	Temperature (°C)
t	Thermal stability time (h)

Subscripts

amb	Ambient
FF	Fresh-food compartment
FZ	Freezer compartment
FZ_1	Heat removal from the initial temperature to the freezing point of thermal load
FZ_2	Heat removal to freeze the thermal load
FZ_3	Heat removal from the freezing point to the final temperature below the freezing point
frost	Referring to ice
water	Referring to water

References

1. Coulomb, D.; Dupont, J.L.; Pichard, A. 29th Informatory note on refrigeration technologies. In *The Role of Refrigeration in the Global Economy*; IIR document; IIR (International Institute of Refrigeration): Paris, France, 2015.
2. Instituto Nacional de Estadística y Geografía (INEGI). Encuesta Nacional de Ingresos y Gastos de los Hogares 2016 Nueva Serie. Available online: http://www.beta.inegi.org.mx/proyectos/enchogares/regulares/enigh/nc/2016/ (accessed on 14 June 2018).
3. General Commission for the Efficient Use of Energy. Secretariat of Energy. 2014. Available online: https://www.gob.mx/conuee#3548 (accessed on 4 September 2018).
4. Bansal, P.K. Developing new test procedures for domestic refrigerators: Harmonization issues and future R&D needs—A review. *Int. J. Refrig.* **2003**, *26*, 735–748.
5. Belman-Flores, J.M.; Barroso-Maldonado, J.M.; Rodríguez-Muñoz, A.P.; Camacho-Vázquez, G. Enhancements in domestic refrigeration, approaching a sustainable refrigerator—A review. *Renew. Sustain. Energy Rev.* **2015**, *51*, 955–968. [CrossRef]
6. Diario Oficial de la Federación (DOF). Norma Oficial Mexicana PROY-NOM-015-ENER-2017, Eficiencia Energética de Refrigeradores y Congeladores Electrodomésticos. Límites, Métodos de Prueba y Etiquetado. Available online: http://www.dof.gob.mx/nota_detalle.php?codigo=5497682&fecha=19/09/2017 (accessed on 18 May 2018).
7. Harrington, L.; Aye, L.; Fuller, B. Impact of room temperature on energy consumption of household refrigerators: Lessons from analysis of field and laboratory data. *Appl. Energy* **2018**, *211*, 346–357. [CrossRef]
8. Masjuki, H.H.; Saidur, R.; Choudhury, I.A.; Mahlia, T.M.I.; Ghani, A.K.; Maleque, M.A. The applicability of ISO household refrigerator–freezer energy test specifications in Malaysia. *Energy* **2001**, *26*, 723–737. [CrossRef]
9. Ozkan, D.B.; Ozil, E.; Inan, C. Experimental investigation of the defrosting process on domestic refrigerator finned tube evaporators. *Heat Transf. Eng.* **2012**, *33*, 548–557. [CrossRef]
10. Harrington, L.; Aye, L.; Fuller, R.J. Opening the door on refrigerator energy consumption: Quantifying the key drivers in the home. *Energy Effic.* **2018**, *11*, 1519–1539. [CrossRef]
11. Saidur, R.; Masjuki, H.H.; Choudhury, I.A. Role of ambient temperature, door opening, thermostat setting position and their combined effect on refrigerator-freezer energy consumption. *Energy Convers. Manag.* **2002**, *43*, 845–854. [CrossRef]
12. Hasanuzzaman, M.; Saidur, R.; Masjuki, H.H. Investigation of energy consumption and energy savings of refrigerator-freezer during open and closed door condition. *J. Appl. Sci.* **2008**, *8*, 1822–1831.
13. Hasanuzzaman, M.; Saidur, R.; Masjuki, H.H. Effects of operating variables on heat transfer and energy consumption of a household refrigerator-freezer during closed door operation. *Energy* **2009**, *34*, 196–198. [CrossRef]
14. Khan, M.I.H.; Afroz, H.M.M.; Rohoman, M.A.; Faruk, M.; Salim, M. Effect of different operating variables on energy consumption of household refrigerator. *Int. J. Energy Eng.* **2013**, *3*, 144–150.
15. Janjic, J.; Katic, V.; Ivanovic, J.; Boskovic, M.; Starcevic, M.; Glamoclija, N.; Baltic, M.Z. Temperatures, cleanliness and food storage practices in domestic refrigerators in Serbia, Belgrade. *Int. J. Consum. Stud.* **2016**, *40*, 276–282. [CrossRef]

16. Geppert, J.; Stamminger, R. Do consumers act in a sustainable way using their refrigerator? The influence of consumer real life behavior on the energy consumption of cooling appliances. *Int. J. Consum. Stud.* **2010**, *34*, 219–227. [CrossRef]

17. Geppert, J.; Stamminger, R. Analysis of effecting factors on domestic refrigerator's energy consumption in use. *Energy Convers. Manag.* **2013**, *76*, 794–800. [CrossRef]

18. James, C.; Onarinde, B.A.; James, S.J. The use and performance of household refrigerators: A review. *Compr. Rev. Food Sci. Food Saf.* **2017**, *19*, 160–174. [CrossRef]

19. Meier, A. Refrigerator energy use in the laboratory and in the field. *Energy Build.* **1995**, *22*, 233–243. [CrossRef]

energies

MDPI

Article

Analysis and Optimization of Exergy Flows inside a Transcritical CO_2 Ejector for Refrigeration, Air Conditioning and Heat Pump Cycles

Sahar Taslimi Taleghani, Mikhail Sorin and Sébastien Poncet *

Department of Mechanical Engineering, Université de Sherbrooke, Sherbrooke, QC J1K2R1, Canada;
Sahar.Taslimi.Taleghani@USherbrooke.ca (S.T.T.); mikhail.v.sorin@usherbrooke.ca (M.S.)
* Correspondence: sebastien.poncet@usherbrooke.ca; Tel.: +1-819-821-8000 # 62150

Received: 13 March 2019; Accepted: 2 May 2019; Published: 4 May 2019

Abstract: In this study, the exergy analysis of a CO_2 (R744) two-phase ejector was performed using a 1D model for both single and double choking conditions. The impact of the back pressure on the exergy destruction and exergy efficiencies was presented to evaluate the exergy performance under different working conditions. The results of two exergy performance criteria (transiting exergy efficiency and Grassmann exergy efficiency) were compared for three modes of an ejector functioning: Double choking, single choking and at the critical point. The behavior of three thermodynamic metrics: Exergy produced, exergy consumed and exergy destruction were evaluated. An important result concerning the ejector's design was the presence of a maximum value of transiting exergy efficiency around the critical point. The impact of the gas cooler and evaporator pressure variations on the different types of exergy, the irreversibilities and the ejector global performance were investigated for a transcritical CO_2 ejector system. It was also shown that the transiting exergy flow had an important effect on the exergy analysis of the system and the Grassmann exergy efficiency was not an appropriate criterion to evaluate a transcritical CO_2 ejector performance.

Keywords: two-phase ejector; CO_2; transcritical system; exergy analysis; irreversibility; transiting exergy

1. Introduction

Carbon dioxide (R744) is an appropriate substitution for synthetic refrigerants in refrigeration, air conditioning and heat pump systems due to its specific features. It is a natural refrigerant that is secure, available and inexpensive. It is non- flammable and non-toxic. It has low global warming potential (GWP) and no impact on the ozone layer. Therefore CO_2 is a promising long-term refrigerant for several heating and cooling applications [1–4].

Furthermore, CO_2 can operate in a transcritical cycle due to its low critical temperature. However, compared to a subcritical cycle, the transcritical CO_2 cycle has lower thermodynamic performance owing to the large exergy destruction of an isenthalpic throttling process from a supercritical to a subcritical state [5]. Among different expansion work recovery devices, the ejector is proposed as a desirable device that enables the use of CO_2 at high heat sink temperatures [6]. An ejector expansion device can replace the throttling valve to decrease the irreversibilities by recovering some part of the expansion work and enhance the cycle's performance. It also increases the suction pressure of the compressor that results in reducing the compressor work. Gay [7] was the first to demonstrate the performance improvement of a transcritical CO_2 cycle by a two-phase ejector.

The one-dimensional and homogeneous two-phase ejector model was first developed by Kornhauser [8] for an R12 refrigerant in the ejector expansion recovery cycle (EERC). The performance enhancement of CO_2 ejector cycles compared to the basic expansion valve cycle have been extensively investigated [9–14]. Zhu et al. [9] experimentally investigated the performance of a transcritical CO_2 ejector heat pump water heater system and reported a 10.3% coefficient of performance (COP) improvement over the corresponding basic cycle. Lucas and Koehler [10] obtained a COP improvement of 17% with maximum ejector efficiencies of 0.22 compared to the maximum COP of the conventional expansion valve cycle. Banasiak et al. [12] carried out an experimental and numerical investigation on a CO_2 heat pump using an optimum ejector geometry and reported the maximum COP increase of 8% in their work compared to a conventional cycle. Boccardi et al. [13] experimentally evaluated the performance of a multi-ejector CO_2 heat pump. An optimal multi-ejector configuration was obtained to maximize the COP. The improvement of COP and heating capacity was reported to be 13.8% and 20%, respectively, for the optimal case at investigated conditions. Elbel [14] observed COP and cooling capacity improvements by up to 7% and 8%, respectively, by adapting an ejector in a conventional cycle.

Although there are numerous literature reviews that present an ejector for CO_2 expansion work recovery, most of the existing works are limited to investigate overall system performance and energy efficiency improvement. However, the evaluation of the second law of thermodynamics is useful to determine the amount and locations of the irreversibilities.

A thermodynamic comparison of the transcritical CO_2 ejector cycle with expansion valve and turbine cycles has been presented by Sarkar [15]. He obtained a 9% exergy efficiency improvement by using an ejector over the usual valve for given operating conditions. Fangtian and Yitai [16] performed the evaluation of COP and exergy destruction for a transcritical CO_2 ejector refrigeration system. An improvement of 30% in COP and a reduction of 25% in exergy destruction were obtained in their analysis compared to the conventional system. Deng et al. [17] reported that the ejector could decrease the total exergy destruction by 23% in a CO_2 transcritical cycle compared to the basic cycle. Zhang and Tian [18] obtained a 45% increase in COP and 43% decrease in the ejector exergy destruction of a transcritical CO_2 ejector refrigeration cycle compared to the basic cycle by an optimized suction nozzle pressure drop (SNPD). A comparative study of different transcritical CO_2 ejector refrigeration cycles was performed under the same cooling capacity by Taslimi et al. [19]. The results showed that EERC has the highest COP and exergy efficiency compared to other cycles. It improves the COP and exergy efficiency by up to 23% and 24%, respectively, compared to the basic throttling cycle. The exergy analysis also implied that the major exergy destruction in EERC occurred in the evaporator (about 33% of the total exergy destruction of the cycle) followed by the compressor (25.5%) and the ejector (24.4%). The second law performance of EERC was investigated theoretically for a two-phase constant area ejector using CO_2 by Ersoy and Bilir Sag [20]. The results indicated that the irreversibility of the ejector system can decrease by 39.1% compared to the basic system at given operating conditions. Gullo et al. [21] implemented an advanced exergy analysis to evaluate the thermodynamic performance of a conventional transcritical R744 booster supermarket refrigerating system at the outdoor temperature of 40 °C and proposed a multi-ejector CO_2 system to improve the system performance. It was concluded that the total exergy destruction can be reduced by about 39% in comparison with the conventional booster system. Bai et al. [22] conducted an advanced exergy analysis to investigate the exergy performance of an ejector expansion transcritical CO_2 refrigeration system. It was found that 43.44% of the total exergy destruction could be avoided by improving the system components. They also showed that the compressor had the largest exergy destruction followed by the ejector, evaporator and gas cooler.

An exergy analysis has been performed for a CO_2 air-to-water heat pump using the multi ejector systems by Boccardi et al. [23]. They confirmed that the throttling irreversibilities can be reduced to 46% by adopting the ejector system. The maximum exergy efficiency improvement by 9% was also reported compared to the basic cycle.

Since the ejector is an important component affecting the thermodynamic performance of the refrigeration, air conditioning and heat pump cycle, the objective of this study is the exergy analysis of a two-phase ejector based on the transiting exergy evaluation. Following the concept of the transiting exergy first introduced by Brodyansky et al. [24], Sorin and Khenich [25] evaluated transiting flows for expansion and compression processes operating above, below and across the ambient temperature. Khennich et al. [26] evaluated the overall transit exergy efficiency as well as the efficiencies within different sections of a single phase ejector for R141b refrigerant.

To the authors' best knowledge, no investigation has been reported yet on the use of the transiting exergy analysis in a CO_2 two-phase ejector with the objective to obtain the effects of different operating conditions on the ejector irreversibilities. Figure 1 shows the performance curve of an ejector. The ejectors may work under double choking or single choking other than the critical point based on the operating conditions. The back pressure that gives the maximum entrainment ratio refers to the critical point. Under double choking conditions, both the primary and the secondary flows are choked and the entrainment ratio is constant while the back pressure decreases. Under single choking conditions, the secondary flow is not choked and the entrainment ratio decreases with increasing the back pressure [27]. In the present paper, first, the exergetic analysis of a CO_2 two-phase ejector is carried out for the critical point as well as for single and double choking conditions and the effect of the back pressure on the amount of the exergy destruction and the values of two types of performance criteria, namely the transiting exergy efficiency and the Grassmann exergy efficiency, is investigated. This analysis helps to understand under which working conditions the ejector performs the best. Second, the effects of the operating conditions (gas cooler and evaporator pressure) are investigated on the exergy efficiency as well as exergy destruction of a two-phase ejector by comparing the transiting and conventional exergy definitions.

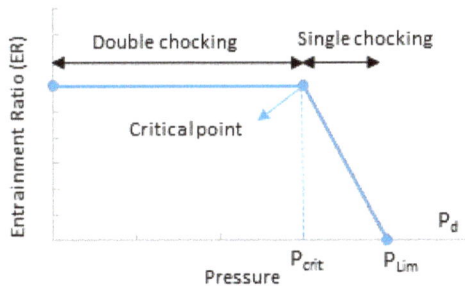

Figure 1. Critical mode of an ejector [28].

2. Theoretical Analysis

A two-phase ejector was used in a transcritical CO_2 cycle to reduce the throttling irreversibilities and improve the cycle efficiency. Figure 3 shows the transcritical CO_2 ejector cycle and its corresponding temperature-specific entropy diagram.

The schematic of an ejector is also shown in Figure 2. A typical ejector comprises a primary nozzle, a secondary nozzle, a mixing section and a diffuser. As shown in Figure 3, the stream at the subcritical state (point 1) is compressed to a supercritical state at high pressure and temperature (point 2). It then releases heat in the gas cooler. The high pressure steam (primary flow) at the gas cooler exit (point 3) expands in the primary nozzle of the ejector into a low pressure and high velocity (point 4). This low pressure entrains the secondary stream into the mixing chamber (point 5). Then two streams mix together (point 6) and the mixture is compressed through the diffuser (point 7) before entering the separator where the two-phase flow is divided to vapor and liquid portions. The vapor portion returns to the compressor while the liquid portion enters the evaporator after expanding through the throttling valve. The secondary stream absorbs heat in the evaporator before entering the ejector.

2.1. Thermodynamic Model

The detailed numerical model of a CO_2 two-phase ejector model for both single choking and double chocking conditions can be found in the authors' previous work [28]. The most important assumptions employed in the model are:

- Flow is one dimensional, steady state and adiabatic in the ejector;
- For two-phase flow, the homogeneous equilibrium model (HEM) is used;
- The thermodynamic and transport properties of CO_2 is based on the real fluid properties;
- The stagnation conditions are assumed at inlets of the primary and secondary flows;
- The friction losses in the nozzles and the diffuser are taken in to account by constant polytropic efficiencies [29] ($\eta_{pol,p} = 0.9$, $\eta_{pol,s} = 0.9$, $\eta_{pol,d} = 0.8$);
- The friction losses in the mixing chamber are neglected, however, a wall friction coefficient is employed to calculate the pressure losses of the constant area part;
- Mass flux maximization criterion is used for choking at the nozzle throats instead of calculating the Mach number due to uncertainty and problematic sound velocity calculations in a two-phase flow;
- Both primary and secondary flows are choked in double choking condition;
- The secondary flow is not choked in single choking condition.

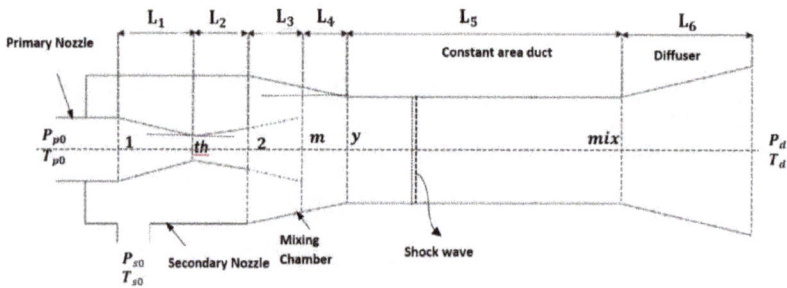

Figure 2. Schematic of an ejector with relevant notations [30].

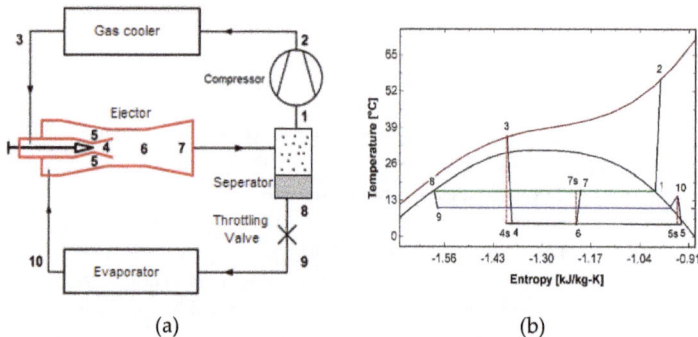

(a)   (b)

Figure 3. (a) Schematic of a transcritical CO_2 ejector cycle and (b) the corresponding temperature-specific entropy diagram [30].

A new methodology was employed here to evaluate the exergy efficiency of a two-phase ejector based on the calculation of the transiting exergy through the ejector under different conditions. The ambient temperature was fixed to 20 °C for exergy calculation. A fixed geometry ejector was used to evaluate its exergy performance under different working conditions. The designed dimensions of the ejector were considered for the given operating conditions ($D_{th} = 0.992$ mm, $D_{mix} = 2.11$ mm, $L_5/D_{mix} = 7.344$, $A_d/A_{mix} = 0.116$) [28].

2.2. Transiting Thermo-Mechanical Exergy in a Two-Phase Ejector

An exergy efficiency definition was used for performance evaluation of a two-phase ejector based on the approach of transiting exergy, presented by Brodyansky et al. [24] that allows non-ambiguous evaluation of two thermodynamic important metrics: Exergy produced and exergy consumed.

$$\eta_{ex,tr} = \frac{\dot{E}_{out} - \dot{E}_{tr}}{\dot{E}_{in} - \dot{E}_{tr}} = \frac{\Delta\dot{E}_{out-tr}}{\nabla\dot{E}_{in-tr}}. \tag{1}$$

where $\nabla\dot{E}$ and $\Delta\dot{E}$ are exergy consumed and produced in the process, $\dot{E}_{in}$ and $\dot{E}_{out}$ are exergy flow rate at the inlet and outlet. It should be mentioned that different terminologies to represent the "exergy produced" and "exergy consumed" in the process are used by some authors. For example, Szargut et al. [31] used "exergy of useful products" vs. "feeding exergy", Kotas [32] used "desired output" vs. "necessary input"; Tsatsaronis [33] and Bejan and Tsatsaronis [34] used "products" vs. "feed".

The difference between the inlet and outlet exergies as well as between the exergy produced and the exergy consumed indicates the exergy destruction (D).

$$D = \nabla\dot{E}_{in,tr} - \Delta\dot{E}_{out,tr} \tag{2}$$

The input-output exergy efficiency, which was first proposed by Grassmann [35], is defined as follows:

$$\eta_{ex,GR} = \frac{\dot{E}_{out}}{\dot{E}_{in}} = 1 - \frac{D}{\dot{E}_{in}} \tag{3}$$

The specific exergy in state k is calculated as:

$$e_k(P,T) = \left[\left(h_k + 0.5u^2\right) - h_0\right] - T_0 \cdot (s_k - s_0)] \tag{4}$$

The specific transiting exergy (e_{tr}) is the lowest exergy value of a material stream, which is defined by the pressure and temperature at the inlet and outlet of a system as well as by the ambient temperature T_0 (in Kelvin). It is illustrated by the following equations:

$$\text{If } (T_{in} > T_0 \text{ and } T_{out} > T_0) : \dot{E}_{tr} = \dot{m}\, e_{tr}(P_{min}, T_{min}, u_{min}) \tag{5}$$

$$\text{If } (T_{in} < T_0 \text{ and } T_{out} < T_0) : \dot{E}_{tr} = \dot{m}\, e_{tr}(P_{min}, T_{max}, u_{min}) \tag{6}$$

$$\text{If } (T_{in} > T_0 \text{ and } T_{out} < T_0) \text{ OR } (T_{in} < T_0 \text{ and } T_{out} > T_0) : \dot{E}_{tr} = \dot{m}\, e_{tr}(P_{min}, T_0, u_{min}) \tag{7}$$

These equations demonstrate that $\dot{E}_{tr}$ is obtained based on the minimum values of the pressure and velocity among the inlet and outlet but it varies for temperature depending on the processes operating in sub-ambient, above ambient or across the ambient temperature.

Let us illustrate the physical meaning of transiting exergy flow on the example of secondary gas flow through a section of an ejector. The Grassmann exergy diagram is presented in Figure 4. Gas enters the section under sub environmental conditions ($T_{in} < T_0$), but leaves it with a temperature higher than environmental ($T_{out} > T_0$); the pressure of the gas is reduced ($P_{out} < P_{in}$), but the velocity rises ($u_{out} > u_{in}$). The exergy values at the inlet and outlet of the section are defined by using the "dead" environmental state as a reference point characterized by the values ($P_0, T_0, u_0 = 0$). According to Equation (7), the transiting exergy is defined by the values (P_{out}, T_0, u_{in}). Thus as illustrated by Figure 4 the transiting exergy is no more than a new reference point to evaluate the exergy consumed ($\nabla\dot{E}$) and exergy produced ($\Delta\dot{E}$). Moreover, the subtraction of $\dot{E}_{tr}$ from the inlet and outlet exergy leads automatically and non-ambiguously to the definition of two terms of exergy consumption and two terms of exergy production. The exergy consumption ($\nabla\dot{E}(P_{in} \rightarrow P_{out}, T_{in} \rightarrow T_0)_{u_{in}}$) is the decrease in thermo-mechanical exergy due to the pressure drop from P_{in} to P_{out} and the temperature rise from T_{in} (sub environmental level) to T_0 at the condition of constant u_{in}. The exergy production $(\Delta \dot{E}(u_{out} \to u_{in}, T_{out} \to T_0)_{P_{out}})$ is the increase in thermo-mechanical exergy due to the velocity rise from u_{in} to u_{out} and the temperature rise from T_0 to T_{out} under the condition of constant pressure P_{out}. The numerical evaluation of exergy consumed and produced for two ejector flows is presented in the following section.

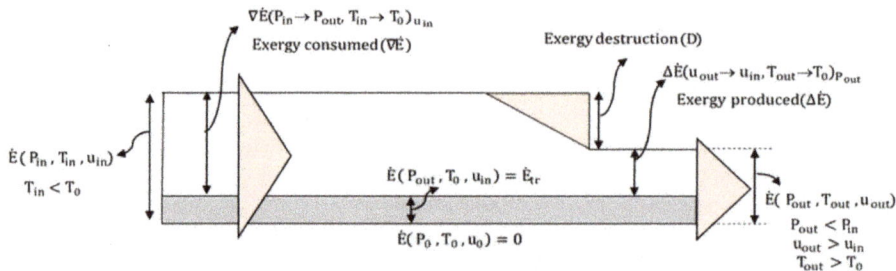

Figure 4. Grassmann diagram with transiting exergy.

2.3. The Exergy Production and Consumption in a Two-Phase Ejector

In a two-phase ejector, the primary stream with high pressure (P_{p0}) and temperature ($T_{p0} > T_0$) expands through the primary nozzle and reaches a low pressure and high-velocity. This supersonic stream entrains the secondary stream at low pressure (P_{s0}) and temperature ($T_{p0} < T_0$) into the mixing section. Inside the mixing section, the two streams exchange momentums and energies and then the mixture compresses to a pressure higher than the secondary inlet pressure ($P_{s0} < P_d < P_{p0}$) and ($T_{s0} < T_d < T_0 < T_{p0}$). The ejector's performance is defined by two parameters: The entrainment ratio (ER $= \dot{m}_s/\dot{m}_p$), which is the ability to entrain the secondary flow inside the ejector and the pressure ratio ($P_{ratio} = P_d/P_{s0}$), the ability to increase the secondary pressure. Figure 5 presents the specific exergy-enthalpy diagram for a two-phase ejector.

The e-h diagram shows the expansion of the primary flow and compression of the secondary flow in the ejector. The primary stream is expanded across T_0 and the secondary stream is compressed at the sub-ambient condition. Equations (8) and (9) were therefore applied to calculate $e_{tr,p}$ and $e_{tr,s}$.

$$e_{tr,p} = e(P_d, T_0) \tag{8}$$

$$e_{tr,s} = e(P_{s0}, T_d) \tag{9}$$

The total exergies consumed and produced by the primary and secondary streams were evaluated by the following equations:

$$\begin{aligned} VE_{p0s0-tr} &= \dot{m}_p\left[e\left(P_{p0}, T_{p0}\right) - e(P_d, T_0)\right] + \dot{m}_s[e(P_{s0}, T_{s0}) - e(P_{s0}, T_d)] \\ &= \dot{m}_p(Ve_{P,T}) + \dot{m}_s \, (Ve_T)_{P_{s0}} \end{aligned} \tag{10}$$

$$\begin{aligned} \Delta E_{d-tr} &= \dot{m}_p[e(P_d, T_d) - e(P_d, T_0)] + \dot{m}_s[e(P_d, T_d) - e(P_{s0}, T_d)] \\ &= \dot{m}_p \, (\Delta e_T)_{P_d} + \dot{m}_s(\Delta e_P)_{T_d} \end{aligned} \tag{11}$$

The exergy consumption and production are linked to both primary and secondary flows. The first term in Equation (10), $(Ve_{P,T})$ is the decrease of the specific thermo-mechanical exergy due to the expansion process and the temperature drop of the primary flow. The second term, $(Ve_T)_{P_{s0}}$ is the decrease of the specific thermal exergy of the secondary flow due to the temperature rise under sub-ambient conditions at constant pressure P_{s0}.

The first term of exergy produced, $(\Delta e_T)_{P_d}$ represents the increase of the specific thermal exergy due to the temperature drop of primary flow from T_0 to T_d under constant pressure P_d. The second term represents the increase in the mechanical exergy component of the secondary flow due to the pressure rise from P_{s0} to P_d at constant temperature T_d.

The main shortcoming of the Grassmann efficiency is the fact that it cannot reveal the real exergy consumption and production within the process. As an example, $(Ve_T)_{P_{s0}}$ represents the exergy consumed due to the temperature rise in the sub-ambient area from T_{s0} to T_d. In fact, this is the partial cold destruction. It means that the cold produced in the evaporator of a refrigeration cycle is destroyed in the ejector. Due to an important transiting exergy flow, the Grassmann exergy efficiency "does not see" this phenomenon. Meanwhile, the transiting exergy definition allows discovering non-ambiguous calculation of exergy consumed and produced and prompts to find the way to recover the amount of the cold destroyed in the ejector.

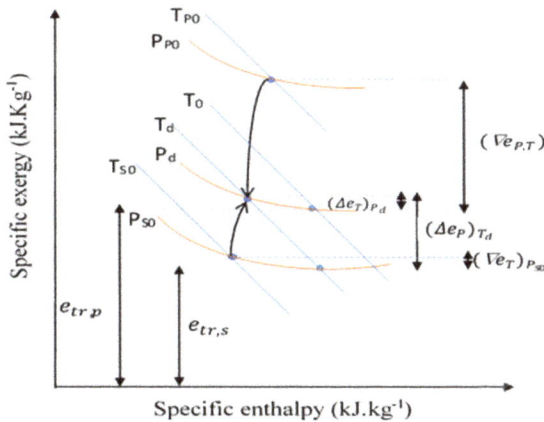

Figure 5. The exergy-enthalpy diagram for expansion and compression processes of a transcritical CO_2 ejector.

3. Results and Discussion

The numerical model for the exergy evaluation of a CO_2 two-phase ejector was presented using the Engineering Equation Solver (EES) software, which is used for the solution of non-linear equations with thermodynamic property functions. This exergy analysis helps to determine the irreversibilities and exergy efficiencies in a two-phase ejector especially when it does not work at its design condition.

3.1. Exergy Analysis of a Fixed Geometry CO_2 Two-Phase Ejector

The calculated parameters for a CO_2 two-phase ejector operating under single choking and double choking conditions, as well as its design condition, are listed in Table 1 [28].

The inlet pressures and temperatures of the primary and secondary flows remained constant for all cases (P_{p0} = 10112 kPa, T_{p0} = 39.3 °C, P_{s0} = 3952 kPa, T_{s0} = 5.5 °C) while the back pressure (diffuser outlet pressure) changed according to the ejector critical conditions.

The first row of Table 1 refers to the critical point of the ejector for a fixed geometry (base case). The second row refers to double choking conditions in which the back pressure is lower than the critical point while the inlet conditions are the same as their design values. The third row presents the single choking conditions in which ER reduces from its critical point (P_{crit}) to maximum limited pressure (P_{lim}; Figure 1).

The values of the numerical calculation for double chocking and single chocking conditions are shown in Tables 2 and 3, respectively. The exergy destruction (D), Grassmann exergy efficiency ($\eta_{ex,GR}$) and transiting exergy efficiency ($\eta_{ex,tr}$) were calculated using equations (1–3). The first rows refer to the results for the base case (critical or design point).

The corresponding exergy produced and exergy consumed of a two-phase ejector were evaluated as well. The exergy analysis based on both transit and Grassmann definitions were also compared and the effect of the ejector's back pressure on the exergy of a two-phase ejector was investigated.

Figure 6 also illustrates the variations of the exergy efficiencies and exergy destruction within the ejector for various back pressures including the critical point, single and double choking conditions.

According to Table 2 and Figure 6, when the back pressure decreased below the critical pressure, the exergy destruction of the ejector increased. The Grassmann exergy efficiency remained approximately constant. The value of $\eta_{ex,GR}$ remained in the range 0.9698–0.9749 (0.5%). While the transiting exergy efficiency ($\eta_{ex,tr}$) decreased by about 5.2%, $\eta_{ex,tr}$ remained within the range 0.564–0.595. The minimum exergy destruction (0.34 kW) took place at the critical back pressure (4601 kPa).

The comparison of the Grassmann and transiting exergy efficiencies showed that $\eta_{ex,tr}$ calculated by Equation (1) was lower than the "optimistic" value given by the Grassmann exergy efficiency. This discrepancy was justified by the presence of transiting exergy flow (e_{tr}), which was neglected when using the Grassmann exergy efficiency. This important result indicates the influence of transiting exergy flow (e_{tr}) inside a two-phase ejector.

Table 1. Calculated parameters of a CO_2 two-phase ejector for different operating conditions.

States	P_d (kPa)	T_d (°C)	$\dot{m}_p$ (kg·s^{-1})	$\dot{m}_s$ (kg·s^{-1})	$\dot{m}_d$ (kg·s^{-1})	P_{ratio}	ER
Base case: (critical point) ($P_d = P_{cr}$)	4601	10.88	0.0423	0.0240	0.0663	1.164	0.568
Double chocking ($P_d < P_{cr}$)	4580	10.61	0.0423	0.0240	0.0663	1.159	0.568
	4520	10.08	0.0423	0.0240	0.0663	1.144	0.568
	4480	9.718	0.0423	0.0240	0.0663	1.134	0.568
	4420	9.162	0.0423	0.0240	0.0663	1.118	0.568
Single choking ($P_d > P_{cr}$)	4730	12.01	0.0423	0.0228	0.0650	1.197	0.539
	4811	12.7	0.0423	0.0211	0.0633	1.217	0.499
	4939	13.77	0.0423	0.0172	0.0595	1.250	0.407
	5002	14.3	0.0423	0.0148	0.0571	1.266	0.351
	5097	15.08	0.0423	0.0108	0.0531	1.290	0.256
	5198	15.89	0.0423	0.0064	0.0486	1.315	0.151
	5313	16.82	0.0423	0.0009	0.0431	1.344	0.021

Another important result derived from the transiting exergy calculation revealed that the exergy produced (Equation (11)) increased by decreasing the back pressure while according to the Grassmann exergy definition, the outlet exergy of the ejector stayed almost constant. However, the increase in exergy production ($\Delta\dot{E}$) was surpassed by the increase in exergy consumption ($\dot{V}\dot{E}$). As a result $\eta_{ex,tr}$ decreased.

Table 3 illustrates the variations of exergy destruction, $\eta_{ex,tr}$ and $\eta_{ex,GR}$ for various back pressures in single choking conditions. It may be observed that the transiting exergy efficiency increased slowly from 0.595 to 0.609 (about 2%) with increasing the back pressure from 4601 kPa (critical point) to 4730 kPa and then decreased to about 73% when the back pressures increased to 5313 kPa. The same justification as the previous part holds. Since both exergy consumed and exergy produced decreased by increasing the back pressure, the decrease in exergy consumption ($\dot{V}\dot{E}$) was surpassed by the decrease in exergy production ($\Delta\dot{E}$), which resulted in reducing the exergy efficiency ($\eta_{ex,tr}$).

The results also showed a minimum value for the exergy destruction. The exergy destruction decreased from its design value, 0.34 kW to 0.28 kW (17%) and then increased to the value of 0.36 kW (higher than that of the critical point) when the back pressure increased to a pressure close to its limited pressure (P_{lim}) 5313 kPa. The comparison of the exergy destruction, ($\eta_{ex,tr}$) and ($\eta_{ex,GR}$) for three different cases (single choking, critical point and double choking) are presented in Figure 7. Two important observations can be made from these results. First is that the minimum value of

η_{exTR} and maximum exergy destruction occurred at single choking mode at maximum pressure (P_{lim}). The maximum value of $\eta_{ex,tr}$ was obtained at the ejector critical point although the exergy destruction was higher at this point compared to some cases of single choking mode.

Table 2. Exergy metrics of a two-phase ejector for different back pressures at double choking conditions ($P_d < P_{cr}$).

	Transiting Exergy Calculation					Grassmann Exergy Calculation		
Back Pressure (P_d, kPa)	Exergy consumed ($V\dot{E}$, kW)	Exergy Produced ($\Delta\dot{E}$, kW)	Exergy destruction (D, kW)	Transiting Exergy ($\dot{E}_{tr}$, kW)	Exergy Efficiency ($\eta_{ex,tr}$)	Exergy Efficiency ($\eta_{ex,GR}$)	Inlet Exergy ($\dot{E}_{in}$, kW)	Outlet Exergy ($\dot{E}_{out}$, kW)
4601	0.839	0.4998	0.34	12.718	0.595	0.975	13.56	13.22
4580	0.852	0.507	0.346	12.705	0.594	0.975	13.56	13.21
4520	0.885	0.52	0.365	12.673	0.587	0.973	13.56	13.19
4480	0.907	0.522	0.385	12.65	0.576	0.972	13.56	13.17
4420	0.941	0.53	0.41	12.617	0.564	0.97	13.56	13.15

Table 3. Exergy metrics of a two-phase ejector for different back pressures and entrainment ratios at single choking conditions (P_d P_{cr}).

	Transit Exergy Calculation					Grassmann Exergy Calculation		
Back Pressure (P_d, kPa)	Exergy consumed ($V\dot{E}$, kW)	Exergy Produced ($\Delta\dot{E}$, kW)	Exergy destruction (D, kW)	Transiting Exergy ($\dot{E}_{tr}$, kW)	Exergy Efficiency ($\eta_{ex,tr}$)	Exergy Efficiency ($\eta_{ex,GR}$)	Inlet Exergy ($\dot{E}_{in}$, kW)	Outlet Exergy ($\dot{E}_{out}$, kW)
4601	0.839	0.5	0.34	12.718	0.595	0.975	13.56	13.22
4730	0.763	0.465	0.298	12.551	0.609	0.978	13.31	13.02
4811	0.711	0.426	0.285	12.276	0.599	0.978	12.99	12.7
4939	0.629	0.348	0.281	11.618	0.553	0.977	12.25	11.97
5002	0.59	0.297	0.293	11.2	0.504	0.975	11.79	11.5
5097	0.533	0.229	0.305	10.495	0.429	0.972	11.03	10.72
5198	0.479	0.155	0.324	9.693	0.324	0.968	10.17	9.848
5313	0.425	0.069	0.356	8.699	0.162	0.961	9.123	8.768

It is important because the design of the ejectors are usually conducted according to the critical point conditions, which leads to a maximum in transiting exergy efficiency, not a minimum of exergy destruction. This is due to the fact that maximum value $\eta_{ex,tr}$ establishes an optimal trade-off between the realization of the ejector's technical purpose (to achieve maximum compression for a given entrainment ratio) and exergy destruction. The second observation is that the Grassmann exergy efficiency did not change with the critical pressure variation, because of important transiting exergy flow. It means that $\eta_{ex,GR}$ was not the appropriate criterion to determine the exergy efficiency of a two-phase ejector.

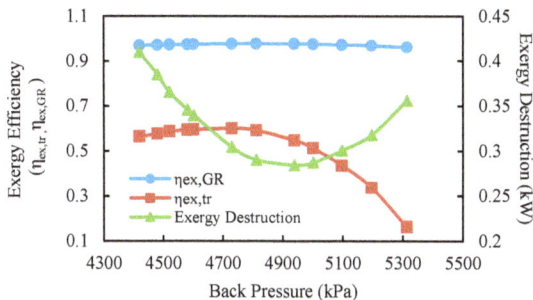

Figure 6. Variations of exergy destruction; transiting exergy efficiency and Grassmann exergy efficiency of the ejector with back pressure.

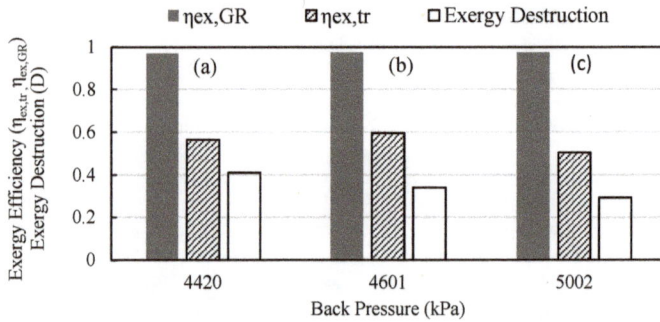

Figure 7. Exergy destruction, transiting exergy efficiency and Grassmann exergy efficiency of the ejector for different conditions: (**a**) Single choking; (**b**) critical point and (**c**) double choking.

3.2. The Comparison of Transiting and Conventional Exergy Evaluation in a Transcritical CO_2 Ejector Cycle

In order to evaluate the effect of transiting exergy to analyze the exergy performance of a two-phase ejector in a cycle, the effects of different operating conditions in a transcritical CO_2 heat pump cycle were investigated.

An ejector heat pump system simulation model developed in the authors' previous work was used for exergy evaluation in this section [30].

As shown in Figures 6 and 7, the ejector had the best performance at its critical conditions. Therefore it is very important that ejector works at its critical conditions. In this section, a transcritical CO_2 cycle using a designed model of the ejector with adjustable throats was used to keep the ejector at its critical conditions (double choking). In this analysis, the simulation results were evaluated for different gas cooler and evaporator pressures. Inlet mass flow rates and temperatures of the external fluid for the gas cooler and evaporator were constant. The gas cooler pressure was in the range of 9000–11,500 kPa and the evaporator pressure was in the range of 2600–4000 kPa. The parameters used for the exergy analysis of the ejector cycle are given in Table 4.

Table 4. Parameters used for the cycle simulation [30].

Parameter	Value	Parameter	Value
P_{gc}, kPa	9000–11,500	D_{th}, mm	1.1–2.6
$T_{gi,ef}$, °C	27.39	D_{mix}, mm	4
$\dot{m}_{gc,ef}$, kg·s^{-1}	0.117	L_5/D_{mix}	8
P_{ev}, kPa	2600–4000	A_{mix}/A_d	0.2
$T_{ei,ef}$, °C	18.04	A_{gc}, m^2	2.199
$\dot{m}_{ev,ex}$, kg·s^{-1}	0.764	A_{ev}, m^2	1.935

3.2.1. The Effect of Gas Cooler Pressure on Exergy Efficiency and Exergy Destruction of the Ejector

The calculated parameters of a CO_2 two-phase ejector operating for different gas cooler pressure are listed in Table 5.

Figure 8 depicts the variation of transiting and Grassmann exergy efficiency with the gas cooler pressure. Table 6 presents the important metrics of two exergy efficiency definitions.

As shown in Figure 8, there existed a maximum exergy efficiency corresponding to an optimum gas cooler pressure. The optimal gas cooler pressure was about 11,000 kPa, which was almost the same as that corresponding to the COP obtained in the authors' previous work [30]. This result shows that the optimal design of a transcritical CO_2 ejector cycle led to system performance improvement in terms of both the first and second laws of thermodynamics.

It was shown that both ejector exergy efficiencies increased when the gas cooler pressure was varied from 9000 kPa to 11,000 kPa. However, the transit exergy efficiency increased by up to 38.3%

while the Grassmann exergy efficiency increased up to 0.85%. Moreover, exergy consumed decreased and exergy produced increased with increasing gas cooler pressure that resulted in an increase of transiting exergy efficiency.

Table 5. Calculated parameters of a CO_2 two-phase ejector for different gas cooler pressures.

$\dot{m}_p$ (kg·s^{-1})	$\dot{m}_s$ (kg·s^{-1})	P_p (kPa)	T_p (°C)	P_s (kPa)	T_s (°C)	P_d (kPa)	T_d (°C)
0.162	0.047	9000	40.53	2780.36	27.05	4189.78	7.13
0.152	0.052	9500	41.63	2780.36	26.92	4067.96	5.96
0.139	0.056	10,000	41.23	2780.36	26.82	3882.57	4.14
0.123	0.059	10,622	38.08	2780.36	26.66	3650.25	1.76
0.107	0.059	11,000	33.18	2780.36	26.62	3405.96	−0.86
0.091	0.056	11,500	26.57	2780.36	26.62	3141.83	−3.86

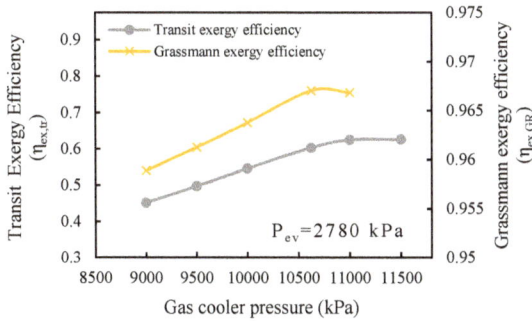

Figure 8. Comparison of two exergy efficiencies as a function of the gas cooler pressure.

Table 6. Exergy metrics of CO_2 two-phase ejector for different gas cooler pressure.

	Transit Exergy Calculation						Grassmann Exergy Calculation		
Gas cooler pressure	Exergy consumed	Exergy produced	Primary transiting Exergy	Secondary transiting Exergy	Transit exergy efficiency	Exergy destruction	Grassmann Exergy efficiency	Inlet exergy	Outlet exergy
(P_{gc}, kPa)	($V\dot{E}$, kW)	($\Delta\dot{E}$, kW)	($\dot{E}_{tr,p}$, kW)	($\dot{E}_{tr,s}$, kW)	($\eta_{ex,tr}$)	(D, kW)	($\eta_{ex,GR}$)	($\dot{E}_{in}$, kW)	($\dot{E}_{out}$, kW)
9000	3.206	1.447	31.354	8.196	0.451	1.758	0.959	42.756	40.997
9500	3.195	1.588	29.228	9.041	0.497	1.607	0.961	41.464	39.856
10,000	3.135	1.711	26.375	9.777	0.546	1.424	0.964	39.287	37.863
10,622	3.019	1.820	22.969	10.398	0.603	1.199	0.967	36.386	35.186
11,000	2.915	1.819	19.814	10.296	0.624	1.095	0.967	33.025	31.930
11,500	2.791	1.746	16.481	9.741	0.625	1.046	0.964	29.013	27.968

3.2.2. The Effect of Evaporator Pressure on Exergy Efficiency and Exergy Destruction of the Ejector

Table 7 presents the calculated operating parameters of a CO_2 two-phase ejector cycle for different evaporator pressure.

Figure 9 and Table 8 illustrate the variation of transiting and Grassmann exergy efficiency for various evaporator pressures.

It can be seen that the transiting exergy efficiency decreased significantly with an increase in evaporator pressure (about 87.8%) while the Grassmann exergy efficiency had a different trend. It decreased slowly at lower evaporator pressure (0.1%) and then increased when the evaporator pressure increased from 3000 kPa to 4000 kPa.

The same as previous results, the range of changes in Grassmann exergy efficiency was very small. However, the important result derived from this analysis dealt with the effect of transit exergy flow inside the ejector. The results showed that Grassmann exergy efficiency had a different trend as compared with transiting exergy efficiency. It can be seen in Table 8, both exergy consumed and exergy produced decreased while the transiting exergies increased as the evaporator pressure increased. However the decrease of exergy consumption (39.6%) was lower than the decrease in exergy production (90.5%), so the transiting exergy efficiency decreased. This result was expected since the ejector exergy destruction increases with the evaporator pressure.

However, according to Grassmann exergy efficiency definition, both inlet and outlet exergy increased by increasing evaporator pressure but the increase in inlet exergy was surpassed by the increase in outlet exergy, which increased Grassmann exergy efficiency.

The main reason for the different trend of exergy efficiencies was the presence of transiting exergy flow that was neglected when the Grassmann exergy efficiency was used. The results also showed that transiting exergy efficiency had a similar trend as the COP when the evaporator pressure increased [30].

Table 7. Calculated parameters of a CO_2 two-phase ejector for different evaporator pressures.

$\dot{m}_p$ (kg·s^{-1})	$\dot{m}_s$ (kg·s^{-1})	P_p (kPa)	T_p (°C)	P_s (kPa)	T_s (°C)	P_d (kPa)	T_d (°C)
0.126	0.054	10,000	39.66	2648.58	26.86	3658.91	1.85
0.139	0.056	10,000	41.23	2780.36	26.82	3882.57	4.14
0.168	0.058	10,000	43.57	3045.77	26.70	4358.38	8.70
0.229	0.059	10,000	46.08	3485.04	26.70	5239.97	16.23
0.297	0.058	10,000	47.05	3969.42	26.70	6001.80	21.99

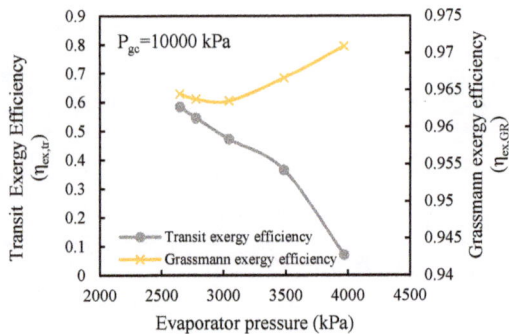

Figure 9. Comparison of two exergy efficiencies as a function of the evaporator pressure.

Table 8. Exergy metrics of CO_2 two-phase ejector for different evaporator pressures.

	Transit Exergy Calculation						Grassmann Exergy Calculation		
Evaporator pressure	Exergy consumed	Exergy produced	Primary transiting Exergy	Secondary transiting Exergy	Transit exergy efficiency	Exergy Destruction	Grassmann Exergy efficiency	Inlet exergy	Outlet exergy
(P_{gc}, kPa)	($\dot{VE}$, kW)	($\Delta\dot{E}$, kW)	($\dot{E}_{tr,p}$, kW)	($\dot{E}_{tr,s}$, kW)	($\eta_{ex,tr}$)	(D, kW)	($\eta_{ex,GR}$)	($\dot{E}_{in}$, kW)	($\dot{E}_{out}$, kW)
2648.58	3.090	1.806	23.667	9.383	0.584	1.285	0.965	36.140	34.855
2780.36	3.135	1.711	26.375	9.777	0.546	1.424	0.964	39.287	37.863
3045.77	3.198	1.510	32.641	10.415	0.472	1.688	0.964	46.254	44.566
3485.04	3.177	1.163	46.184	10.924	0.366	2.014	0.967	60.285	58.271
3969.42	2.354	0.166	61.658	11.146	0.071	2.188	0.971	75.158	72.970

4. Conclusions

An exergy analysis based on the transiting exergy was employed to evaluate the exergy destruction and exergy efficiency of a CO_2 two-phase ejector at its critical point as well as under double choking and single choking conditions. The results were compared with the conventional Grassmann exergy analysis. This application provided the evaluation of exergy destruction as well as useful exergy production in the ejector. Two important thermodynamic metrics, exergy produced and exergy consumed were obtained for different ejector working conditions. It also provided the information regarding the transit exergy flows in a two-phase ejector that cannot be obtained through the conventional exergy analysis.

There was a compromise between exergy destruction and useful exergy produced in the ejector to indicate its performance, which cannot be derived from the Grassmann exergy analysis. The transiting exergy efficiency achieved the maximum value at the critical pressure corresponding to the critical point, it confirms a well-known heuristics, to design ejectors according to the conditions of the critical point. The impact of the gas cooler and evaporator pressures was investigated on ejector exergy efficiency in a transcritical CO_2 cycle. The exergy efficiency had a different trend as a function of evaporator pressure when evaluated by the transiting or Grassmann exergy definition.

The established results showed that the transiting exergy flow had an important effect on the ejector exergy performance. On the contrary, the Grassmann exergy efficiency was not an appropriate criterion for the exergy evaluation of a two-phase ejector. The approach based on transiting exergy definition provided useful information, which can be used for the improvement of the ejector systems.

Future work would involve the investigation of the transiting exergy flow on advanced exergy analysis of a transcritical CO_2 cycle.

Author Contributions: S.T.T., principal investigator, performed the numerical simulations of the ejector and heat pump system. M.S. proposed the concept of transiting exergy and supervised the findings of this work. S.P. provided technical and scientific assistance. All authors contributed to the final manuscript.

Funding: This research received no external funding.

Acknowledgments: This project is a part of the Collaborative Research and Development (CRD) Grants Program at "Université de Sherbrooke". The authors acknowledge the support of the Natural Sciences and Engineering Research Council of Canada, Hydro-Québec, Rio Tinto Alcan and Canmet ENERGY Research Center of Natural Resources Canada (RDCPJ451917-13).

Conflicts of Interest: The authors declare no conflict of interest.

Nomenclature

A	Cross section area, mm^2
D	Diameter, mm
$\dot{E}$	Exergy rate, kW
e	Specific exergy, kJ·kg^{-1}
ER	Entrainment ratio
h	Specific enthalpy, kJ·kg^{-1}
L	Length, m
$\dot{m}$	Mass flow rate, kg·s^{-1}
P	Pressure, kPa
P_d	Back pressure (discharge pressure), kPa
P_{ratio}	Pressure ratio
s	Specific entropy, kJ·kg^{-1}·K^{-1}
T	Temperature, °C
u	Mean axial velocity, m·s^{-1}

Greek symbols

η	Efficiency
∇	Consumption
Δ	Production

Subscripts and superscripts

0	Ambient state
crit	Critical
d	Diffuser outlet
ev	Evaporator
ei	Evaporator inlet
ex	Exergy
gc	Gas cooler
gi	Gas cooler inlet
in	Inlet
lim	Limiting
mix	Mixing
out	Outlet
p	Primary
pol	Polytropic
s	Secondary
ef	External fluid
th	Ejector's throat
tr	Transiting

Abbreviations

COP	Coefficient of performance
EERC	Ejector expansion recovery cycle
EES	Engineering equation solver
GWP	Global warming potential
HEM	Homogeneous equilibrium model
SNPD	Suction nozzle pressure drop

References

1. Kim, M.H.; Pettersen, J.; Bullard, C.W. Fundamental process and system design issues in CO_2 vapor compression systems. *Prog. Energy Combust. Sci.* **2004**, *30*, 119–174. [CrossRef]
2. Lorentzen, G. Revival of carbon dioxide as a refrigerant. *Int. J. Refrig.* **1994**, *17*, 292–301. [CrossRef]
3. Gullo, P.; Hafner, A.; Banasiak, K. Transcritical R744 refrigeration systems for supermarket applications: Current status and future perspectives. *Int. J. Refrig.* **2018**, *93*, 269–310. [CrossRef]
4. Austin, B.T.; Sumathy, K. Transcritical carbon dioxide heat pump systems: A review. *Renew. Sustain. Energy Rev.* **2011**, *15*, 4013–4029. [CrossRef]
5. Fazelpour, F.; Morosuk, T. Exergoeconomic analysis of carbon dioxide transcritical refrigeration machines. *Int. J. Refrig.* **2014**, *38*, 128–139. [CrossRef]
6. Elbel, S.; Lawrence, N. Review of recent developments in advanced ejector technology. *Int. J. Refrig.* **2016**, *62*, 1–18. [CrossRef]
7. Gay, N.H. Refrigerating system. Google Patents, U.S. Patent No. 1,836,318, 15 December 1931.
8. Kornhauser, A.A. The use of an ejector as a refrigerant expander. In Proceedings of the 1990 USNC/IIR—Purdue Refrigeration Conference, Purdue University, West Lafayette, IN, USA, 17–20 July 1990; pp. 10–19.
9. Zhu, Y.; Huang, Y.; Li, C.; Zhang, F.; Jiang, P.X. Experimental investigation on the performance of transcritical CO_2 ejector–expansion heat pump water heater system. *Energy Convers. Manag.* **2018**, *167*, 147–155. [CrossRef]
10. Lucas, C.; Koehler, J. Experimental investigation of the COP improvement of a refrigeration cycle by use of an ejector. *Int. J. Refrig.* **2012**, *35*, 1595–1603. [CrossRef]
11. He, Y.; Deng, J.; Zheng, L.; Zhang, Z. Performance optimization of a transcritical CO_2 refrigeration system using a controlled ejector. *Int. J. Refrig.* **2017**, *75*, 250–261. [CrossRef]

12. Banasiak, K.; Hafner, A.; Andresen, T. Experimental and numerical investigation of the influence of the two-phase ejector geometry on the performance of the R744 heat pump. *Int. J. Refrig.* **2012**, *35*, 1617–1625. [CrossRef]

13. Boccardi, G.; Botticella, F.; Lillo, G.; Mastrullo, R.; Mauro, A.W.; Trinchieri, R. Experimental investigation on the performance of a transcritical CO_2 heat pump with multi-ejector expansion system. *Int. J. Refrig.* **2017**, *82*, 389–400. [CrossRef]

14. Elbel, S. Historical and present developments of ejector refrigeration systems with emphasis on transcritical carbon dioxide air-conditioning applications. *Int. J. Refrig.* **2011**, *34*, 1545–1561. [CrossRef]

15. Sarkar, J. Optimization of ejector-expansion transcritical CO_2 heat pump cycle. *Energy* **2008**, *33*, 1399–1406. [CrossRef]

16. Fangtian, S.; Yitai, M. Thermodynamic analysis of transcritical CO_2 refrigeration cycle with an ejector. *Appl. Therm. Eng.* **2011**, *31*, 1184–1189. [CrossRef]

17. Deng, J.; Jiang, P.; Lu, T.; Lu, W. Particular characteristics of transcritical CO_2 refrigeration cycle with an ejector. *Appl. Therm. Eng.* **2007**, *27*, 381–388. [CrossRef]

18. Zhang, Z.; Tian, L. Effect of suction nozzle pressure drop on the performance of an ejector-expansion transcritical CO_2 refrigeration cycle. *Entropy* **2014**, *16*, 4309–4321. [CrossRef]

19. Taslimi Taleghani, S.; Sorin, M.; Poncet, S. Energy and exergy efficiencies of different configurations of the ejector-based CO_2 refrigeration systems. *Int. J. Energy Prod. Manag.* **2018**, *3*, 22–33.

20. Ersoy, H.K.; Bilir Sag, N. Performance characteristics of ejector expander transcritical CO_2 refrigeration cycle. *Proc. Inst. Mech. Eng. Part. J. Power Energy* **2012**, *226*, 623–635. [CrossRef]

21. Gullo, P.; Hafner, A.; Banasiak, K. Thermodynamic Performance Investigation of Commercial R744 Booster Refrigeration Plants Based on Advanced Exergy Analysis. *Energies* **2019**, *12*, 354. [CrossRef]

22. Bai, T.; Yu, J.; Yan, G. Advanced exergy analyses of an ejector expansion transcritical CO_2 refrigeration system. *Energy Convers. Manag.* **2016**, *126*, 850–861. [CrossRef]

23. Boccardi, G.; Botticella, F.; Lillo, G.; Mastrullo, R.; Mauro, A.W.; Trinchieri, R. Thermodynamic Analysis of a Multi-Ejector, CO_2, Air-To-Water Heat Pump System. *Energy Procedia* **2016**, *101*, 846–853. [CrossRef]

24. Brodyansky, V.M.; Sorin, M.; Le Goff, P. *The efficiency of industrial processes: Exergy analysis and optimization*; Elsevier Science B. V.: Amsterdam, The Netherlands, 1994.

25. Sorin, M.; Khennich, M. Exergy Flows Inside Expansion and Compression Devices Operating below and across Ambient Temperature. In *Energy Systems and Environment*; IntechOpen: London, UK, 2018.

26. Khennich, M.; Sorin, M.; Galanis, N. Exergy flows inside a one phase ejector for refrigeration systems. *Energies* **2016**, *9*, 212. [CrossRef]

27. Haghparast, P.; Sorin, M.V.; Nesreddine, H. The impact of internal ejector working characteristics and geometry on the performance of a refrigeration cycle. *Energy* **2018**, *162*, 728–743. [CrossRef]

28. Taslimi Taleghani, S.; Sorin, M.; Poncet, S. Modeling of two-phase transcritical CO_2 ejectors for on-design and off-design conditions. *Int. J. Refrig.* **2017**, *87*, 91–105. [CrossRef]

29. Haghparast, P.; Sorin, M.; Nesreddine, H. Effects of component polytropic efficiencies on the dimensions of monophasic ejectors. *Energy Convers. Manag.* **2018**, *162*, 251–263. [CrossRef]

30. Taslimi Taleghani, S.; Sorin, M.; Poncet, S.; Nesreddine, H. Performance investigation of a two-phase transcritical CO_2 ejector heat pump system. *Energy Convers. Manag.* **2019**, *185*, 442–454. [CrossRef]

31. Szargut, J.; Morris, D.R.; Steward, F.R. *Exergy Analysis of Thermal, Chemical, and Metallurgical Processes*; Hemisphere Publ. Corp: New York, NY, USA, 1988.

32. Kotas, T.J. *The Exergy Method of Thermal Plant Analysis*, 2nd ed.; Krieger Publishing: Malabar, FL, USA, 1995.

33. Tsatsaronis, G. Thermoeconomic analysis and optimization of energy systems. *Prog. Energy Combust. Sci.* **1993**, *19*, 227–257. [CrossRef]

34. Bejan, A.; Tsatsaronis, G. *Thermal Design and Optimization*; John Wiley & Sons: New York, NY, USA, 1996.

35. Grassmann, P. Towards the general definition of efficiency (in German). *Chem. Ing. Tech.* **1950**, *22*, 77–80. [CrossRef]

energies

MDPI

Article

Evaluating Magnetocaloric Effect in Magnetocaloric Materials: A Novel Approach Based on Indirect Measurements Using Artificial Neural Networks

Angelo Maiorino [1],*, Manuel Gesù Del Duca [1], Jaka Tušek [2], Urban Tomc [2], Andrej Kitanovski [2] and Ciro Aprea [1]

[1] Department of Industrial Engineering, Università di Salerno, Via Giovanni Paolo II, 132, 84084 Fisciano, Salerno, Italy; mdelduca@unisa.it (M.G.D.D.); aprea@unisa.it (C.A.)
[2] Faculty of Mechanical Engineering, University of Ljubljana, Aškerčeva 6, 1000 Ljubljana, Slovenia; Jaka.Tusek@fs.uni-lj.si (J.T.); urban.tomc@fs.uni-lj.si (U.T.); andrej.kitanovski@fs.uni-lj.si (A.K.)
* Correspondence: amaiorino@unisa.it; Tel.: +390-8996-4105

Received: 10 April 2019; Accepted: 10 May 2019; Published: 16 May 2019

Abstract: The thermodynamic characterisation of magnetocaloric materials is an essential task when evaluating the performance of a cooling process based on the magnetocaloric effect and its application in a magnetic refrigeration cycle. Several methods for the characterisation of magnetocaloric materials and their thermodynamic properties are available in the literature. These can be generally divided into theoretical and experimental methods. The experimental methods can be further divided into direct and indirect methods. In this paper, a new procedure based on an artificial neural network to predict the thermodynamic properties of magnetocaloric materials is reported. The results show that the procedure provides highly accurate predictions of both the isothermal entropy and the adiabatic temperature change for two different groups of magnetocaloric materials that were used to validate the procedure. In comparison with the commonly used techniques, such as the mean field theory or the interpolation of experimental data, this procedure provides highly accurate, time-effective predictions with the input of a small amount of experimental data. Furthermore, this procedure opens up the possibility to speed up the characterisation of new magnetocaloric materials by reducing the time required for experiments.

Keywords: magnetic refrigeration; magnetocaloric effect; $LaFe_{13-x-y}Co_xSi_y$; gadolinium; artificial neural network; modelling

---

1. Introduction

Over the past two decades many research efforts have been focused on the development of not-in-kind refrigeration technologies [1], defined as alternative options to vapour-compression refrigeration systems. Among these, magnetic refrigeration shows promising results in terms of energy efficiency [1–5]. Furthermore, the scientific community is very interested because of the favourable environmental characteristics of this technology [6,7], which uses a solid substance as the refrigerant rather than the greenhouse-effect-promoting refrigerants used in vapour-compression applications. These latter refrigerants are being phased out [8,9], and they will need to be replaced. However, only a few pure fluids possess the combination of properties necessary for a refrigerant. Unfortunately, these fluids are at least slightly flammable [10]. In addition, some natural fluids, such as isobutane, propane or carbon dioxide, have been proposed as a solution. But, due to their flammability, hydrocarbons can only be employed in countries that permit their use [11]. On the other hand, the use of carbon dioxide is limited since it needs a high working pressure and this leads to poor performance in terms of energy [12,13].

The functioning of a magnetic refrigerator system is based on the magnetocaloric effect (MCE), the discovery of which is attributed to Weiss and Piccard [14]. The MCE is a feature of some magnetic materials that heat up when they are subjected to an external magnetic field, since the atoms, which act as magnetic dipoles, align with the magnetic field, and then cool down after the applied field is removed. They are called magnetocaloric materials (MCMs). The MCE is characterised by the directly measured temperature change of the MCM subjected to the external magnetic field. It can be evaluated as the isothermal entropy change Δs_{iso} or the adiabatic temperature change ΔT_{ad} of the MCM induced by the increasing or the decreasing of an external magnetic field that the material is subjected to. If the conditions are kept adiabatic, the temperature of the MCM increases/decreases by the amount ΔT_{ad}. On the other hand, if the conditions are kept isothermal, the specific entropy of the MCM decreases/increases by the amount Δs_{iso}.

One of the main problems associated with MCMs is related to the small MCE [15], which is not large enough to reach an appropriate temperature span for near-room-temperature applications, i.e., between 20 °C and 35 °C. A theoretical study about the maximum MCE achievable with an MCM, referring to a single-stage cooling device, was presented by Zverev et al. [16]. For achieving a larger temperature span, the prototypes built so far have been based on the active magnetic regenerative (AMR) cycle [17], which makes it possible to increase the temperature span between the heat sink and the heat source, so that it can be several times larger than the adiabatic temperature changes in an MCM. The AMR cycle is based on four operational steps (in the case of the Brayton thermodynamic cycle): an adiabatic magnetisation, isofield cooling, adiabatic demagnetisation and isofield heating. During the first step, the MCM is subjected to the external magnetic field and the temperature of the MCM increases due to the MCE. Then, keeping the magnetic field at a constant value, a fluid can flow through the material absorbing heat from it, which is subsequently rejected in the hot heat exchanger. In the adiabatic demagnetisation step, the external magnetic field is removed, and the MCM cools down. During the last step of the cooling cycle, with no external magnetic field, the fluid flows in a counter-flow direction through the material, expelling heat to it. Next, the fluid absorbs heat in the cold heat exchanger. The steps are then continuously repeated.

For room-temperature applications, starting from the construction of the first magnetic refrigerator prototype [18], several devices with different configurations and different magnetocaloric materials have been developed. In these prototypes, Gd and Gd-based alloys were the most commonly used MCMs [19–25]. Nevertheless, other substances have also been tested to evaluate their performances in magnetic refrigeration systems. These include La-Fe-Co-Si [26–30] and also Mn-Fe-P-As [31,32] alloys. Considering low-temperature applications, a comprehensive review about MCMs and devices for magnetic refrigeration in the temperature range of nitrogen and hydrogen liquefaction was performed by Zhang et al. [33]. Furthermore, a very recent review about magnetic refrigerator devices for room-temperature applications can be found in Gimaev et al. [34]. In addition to the experimental investigations, several AMR numerical models have been developed over the years, with the aim of studying different operating conditions and different MCMs in order to understand the feasibility of a magnetic refrigerator application [35,36].

The magnetocaloric properties required for AMR modelling, such as the adiabatic temperature change, can be obtained experimentally by employing direct and indirect methods [37–39]. The former is based on measurements of the temperature of the MCM sample subjected to an external magnetic field, thereby obtaining the adiabatic temperature change directly. The latter is based on heat-capacity and/or magnetisation measurements. With the indirect methods there are two routes to obtaining the magnetisation curves, depending on the temperature and the external magnetic field: the isofield process and the isothermal process. The isothermal measurement is the commonly used technique since it is faster. Hence, once the $M(T, H)$ curves for the different temperatures and external magnetic fields are obtained, it is possible to calculate the isothermal entropy change and the adiabatic temperature change indirectly using Maxwell's relations [37]. Another experimental method has been presented by

Nielsen et al. [40], where the authors show an experimental device that can perform direct measurements of the magnetic entropy change.

Another approach to evaluating the adiabatic temperature change of MCMs is the application of theoretical models, which can be divided into two groups [41]: thermodynamic models and first-principle models. In particular, the formers are frequently used in the area of magnetic refrigeration to evaluate the magnetocaloric effect within numerical models of AMR cycles [42–45]. They identify a link among the magnetisation, the temperature and the external magnetic field using equations of state [41]. The most widely applied thermodynamic model for the calculation of the magnetocaloric effect is the Weiss mean field theory (MFT), which can be used to evaluate the total entropy of a ferromagnetic material as a function of the temperature and the external magnetic field [46] by considering the link with the magnetisation. The first-principle models are based on a calculation of the exchange-coupling energies and the magnetic moments of MCMs. They are usually performed to extract information about the values of magnetization at finite temperatures, the magnetic entropy, the magneto-structural transition temperatures, and the MCE, as shown in Paudyal et al. [47].

However, both the theoretical and experimental methods for evaluating the MCE have some important disadvantages. For example, the theoretical model (such as MFT) requires several material-based data that are not always known (in particular for new MCMs) and usually quite significantly over-predicts the MCE, as shown, for example, in Mugica Guerrero et al. [48]. On the other hand, the experimental methods are time-consuming and could be challenging to perform, especially in the case of direct measurements, where a high density of the data is required (as in the case of AMR modelling). Furthermore, it should be noted that there is no standardisation of the method for evaluating the magnetocaloric properties to characterise MCMs, which does not allow a direct comparison of the results. However, a correct and rapid evaluation of the magnetocaloric properties of an MCM, especially the adiabatic temperature change, and also isothermal entropy change and specific heat, is crucial to understanding the feasibility of using it in a magnetic refrigerator application.

In this study we propose a new MCMs-modelling method based on artificial neural networks (ANNs), which have been used in the past only for modelling the performance of AMR-based refrigerator devices [49,50]. This approach could represent a unique standardised method to evaluate the magnetocaloric effect and make it possible to reduce the experimental efforts to characterise known or new MCMs. A successful ANN can be used in the research field to produce results close to the experimental ones in a much faster way, providing sets of data which can be used for testing more detailed theories. Several mathematical models stemming from machine-learning algorithms were already applied with this purpose, such as the genetic algorithm [51], Bayesian approach [52] and ANNs themselves [53]. Furthermore, the ANN-based procedure can facilitate the implementation of the magnetocaloric properties of MCMs within an AMR numerical model. The latter represents the main advantage of this technique. Indeed, it is only necessary to perform the experiments at a few magnetic fields and temperatures to obtain a mathematical model of the magnetocaloric properties of the MCMs that can be easily included in an AMR numerical model. Then, the proposed ANN model can predict accurately the behaviour of the MCM for any magnetic fields and temperatures in between, ensuring a high density of the data needed for accurate AMR numerical simulations.

2. Materials and Methods

The ANN-based method is divided into four steps starting with the experimental phase that involves isothermal magnetisation and specific heat measurements at different magnetic fields and different absolute temperatures. Then, the collected data are processed to feed the development and the training of the ANN.

The ANN is a mathematical model inspired by the biological neural network in the human brain. An ANN is composed of several simple processing units, named neurons, which are connected to each other through weighted links, i.e., the synaptic weights. A neuron is characterised by three essential elements: a set of synaptic weights, a summation junction and an activation function. The first of these represents the strength of the relationships among adjacent neurons. The summation junction computes a weighted sum of the input and the activation function determines the output of the neuron itself. The main advantage of this model is related to its simplicity and ability to identify complex relationships between the input and the output using experimental data, without requiring any specific equation [50].

In order to define the ANN, it is necessary to specify the number of inputs, its architecture (i.e., the number of layers and the topology), the activation function of each layer and the training algorithm through which the knowledge-extraction process from the experimentations is run, modifying the free parameters of the network (synaptic weights). This task is accomplished by a training phase, during which the synaptic weights are modified to reduce the estimation error of the network. After the learning process, the ANN can predict the magnetisation and the specific heat of an MCM at each magnetic field and each absolute temperature within the range of the training dataset.

The third step foresees the calculation of the isothermal entropy change of the magnetocaloric material using the parameters of the ANN, which are the synaptic weights. In the last step, the adiabatic temperature change is evaluated by the construction of the s-T diagram, with the isothermal entropy-change values calculated in the previous step.

2.1. Experimental Setup and Data Collection

The first step of the ANN-based procedure is the experimental characterisation of the isothermal magnetisation and specific heat data. Two groups of materials were used to test the procedure: commercial gadolinium (Gd) and three different samples of $LaFe_{13-x-y}Co_xSi_y$ (hereto referred to as La-Fe-Co-Si) with exact compositions of ($x = 0.86$, $y = 1.08$), ($x = 0.94$, $y = 1.01$) and ($x = 0.97$, $y = 1.07$) and Curie temperatures of about 276 K, 287 K and 289 K, respectively. They are named Specimen 1, Specimen 2 and Specimen 3, respectively. For the purposes of this work the required experimental data were obtained from Bjørk et al. [26] The magnetisation measurements were performed with a vibrating-sample magnetometer (VSM), while the specific-heat data were collected with a differential scanning calorimeter (DSC) equipped with a magnetic field source. Furthermore, data relating to the adiabatic temperature change, measured with a type-E thermocouple (± 0.1 K), were used to evaluate the performance of the entire ANN-based procedure. The experimental equipment and the procedures for the different tests are explained and presented in detail in Bjørk et al. [26] and Jeppesen et al. [54] It is important to note that all the measured values of the magnetocaloric properties were subsequently evaluated as a function of the internal magnetic field $\mu_0 H_{int}$, which depends on the geometry of the sample (and the demagnetisation factor). Therefore, the thermodynamic properties were obtained as a function of $\mu_0 H_{int}$. The latter is fundamental since the internal field $\mu_0 H_{int}$ (subsequently referred to as H) is used as the input for the ANN in the second step. The output of this step is represented by the magnetisation and the specific heat experimental data of the MCMs, organised in matrix format (Equations (1) and (2)), where the values of the magnetic field H and the absolute temperature T are reported as rows and columns, respectively

$$M = \begin{bmatrix} [] & T_1 & T_2 & \cdots & T_v \\ H_1 & M_{11} & M_{12} & \cdots & M_{1v} \\ H_1 & M_{21} & M_{22} & \ddots & \vdots \\ \vdots & \vdots & \ddots & \ddots & \vdots \\ H_u & M_{u1} & M_{u2} & \cdots & M_{uv} \end{bmatrix}, \tag{1}$$

$$
c_{P,H} = \begin{bmatrix}
[] & T_1 & T_2 & \cdots & T_v \\
H_1 & c_{P,H_{11}} & c_{P,H_{12}} & \cdots & c_{P,H_{1v}} \\
H_1 & c_{P,H_{21}} & c_{P,H_{22}} & \ddots & \vdots \\
\vdots & \vdots & \ddots & \ddots & \vdots \\
H_u & c_{P,H_{u1}} & c_{P,H_{u2}} & \cdots & c_{P,H_{uv}}
\end{bmatrix}.
\tag{2}
$$

2.2. Training the Artificial Neural Network

A multi-layer perceptron (MLP) with two inputs, one hidden layer and two outputs was used as part of the procedure to predict the behaviour of the MCMs (Figure 1).

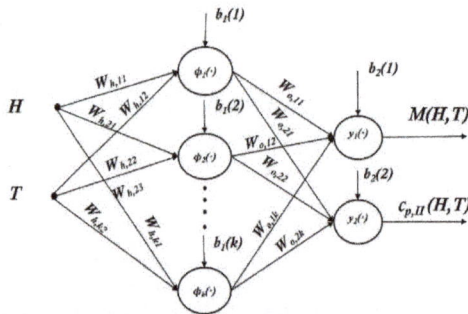

Figure 1. The adopted structure of the ANN. The magnetic field, in *T*, and the absolute temperature, in K, are the inputs of the ANN. The model provides the values of magnetisation, in A/m, and specific heat, in J/kgK, at the given magnetic field and absolute temperature. The subscript h refers to the synaptic weights between the hidden and the input layer. The subscript o refers to the synaptic weights between the output and the hidden layer. The index and the subscript *k* refer to the *k*-th neuron of the hidden layer.

The number of hidden layers can be greater than one, and it depends on the complexity of the problem. For very complex problems, such as vision and human language understanding, ANNs with more than one hidden layer (deep neural networks) can provide better performance [55], but considering approximation problems of continuous functions only one layer is sufficient to obtain good results [56]. If the approximation problem concerns non-continuous function, it may be necessary to use more than one hidden layer. In the framework of MCMs, this could happen considering first-order magnetic transition materials, characterised by a discontinuity in the magnetisation. The following equation describes the generalised mathematical model of an MLP with one hidden layer:

$$
y_j = \varphi_o \left(\sum_{k=1}^{N_h} w_{jk} * \varphi_h \left(\sum_{i=1}^{N_{il}} w_{ki} * x_i + b_k \right) + b_j \right),
\tag{3}
$$

where:

- y_j is the output estimated *j* by the ANN;
- φ_o is the output-layer activation function;
- N_h is the number of hidden neurons;
- w_{jk} are the synaptic weights between the output *j* and the hidden neuron *k*;
- φ_h is the hidden-layer activation function;
- N_{il} is the number of inputs;
- w_{ki} are the synaptic weights between the hidden neuron *k* and the input *i*;
- x_i is the input *i* to the ANN;

- b_k, b_j are the bias of the hidden neuron k and the output j, respectively.

Considering the indirect method based on magnetisation measurements, the selected inputs for the model are represented by the magnetic field H, corrected for the demagnetisation factor, and the absolute temperature T. They are re-arranged in a matrix $I^{2 \times p}$, where p is the number of available experimental data, like the following:

$$I^{2 \times p} = \begin{bmatrix} H_1 & H_2 & \cdots & H_p \\ T_1 & T_2 & \cdots & T_p \end{bmatrix}. \tag{4}$$

As outputs, this architecture calculates the value of the magnetisation $M(H,T)$ and the specific heat at a constant magnetic field $c_{P,H}(H,T)$, corresponding to the magnetic fields and temperatures given. For each p, the ANN draws up the inputs according to the synaptic weights and provides the following output matrix:

$$O^{2 \times p} = \begin{bmatrix} M_1 & M_2 & \cdots & M_p \\ c_{P,H_1} & c_{P,H_2} & \cdots & c_{P,H_p} \end{bmatrix}. \tag{5}$$

The number of hidden neurons can be identified by employing a trial-and-error procedure [57,58] or using some empirical rules, as reported in [59]. There is no specific process to evaluate the optimal number of hidden neurons, but it must be identified case by case. In this study the evaluation of the number of hidden neurons was made by employing an iterative trial-and-error process. The minimum number of hidden neurons $N_{h,min}$ was identified using the following empirical rule [59,60]:

$$N_{h,min} = 2 * N_i + 1, \tag{6}$$

where N_i is the number of input neurons. Since the ANN has two inputs, the minimum number of hidden neurons was fixed at 5. Several ANNs were trained to vary the number of hidden neurons between the minimum and the selected maximum value. The latter was fixed at 15 units, but it can be changed to extend the range of the investigation. The activation functions selected for the hidden and output layers are the hyperbolic tangent (Equation (7)) and the linear one (Equation (8)), respectively:

$$\varphi_H(v_k) = \tanh(v_k) = \frac{2}{(1 + e^{-2*v_k})} - 1, \text{ where } v_k = \sum_{i=1}^{N_i} w_{ki} * x_i + b_k, \tag{7}$$

$$\varphi_O(v_j) = v_j, \text{ where } v_j = \sum_{k=1}^{N_h} w_{jk} * H_k + b_j. \tag{8}$$

In Equations (7) and (8), the subscripts k and j refer to the k-th hidden neuron and the j-th output neuron. The term v represents the induced local field of the neuron, which is the weighted sum of its inputs. The ANNs were trained using the standard error back-propagation (EBP) algorithm [61] with cross-validation [62], developed and performed with a code written in the MATLAB environment. The EBP algorithm can be divided into two steps: the forward pass and the backward pass. During the first one, the output of the ANN, fed with the input array $(x_1, x_2, \ldots, x_N)$, is calculated and then the error e_j is evaluated in comparison to the target. This error is used to compute the correction Δw_{jk} of the synaptic-weight values of the output layer. In the backward pass, the error is propagated towards the input layer, and the adjustment Δw_{ki} of the synaptic-weight values of the hidden layer is calculated. The evaluation of the synaptic-weight changes is usually performed according to the steepest descent method [63] and can be expressed using the following equation:

$$\Delta w_{jk} = -\eta * \frac{\partial E(e_j)}{\partial w_{jk}}, \tag{9}$$

where η is the *learning rate*, and $E(e_j)$ is the error function. The error derivative of Equation (9) can be easily calculated through the derivative chain rule [61]. The learning rate for this application has been fixed at 0.3 after a trial-and-error procedure, but it can be modified according to needs. The batch-training method was used to perform the EBP. According to this learning technique, the synaptic weights are updated only when all the examples are fed into the ANN. The range between the processing of the first and the last experimental sample is named epoch. At the end of each epoch, the error metric is calculated, and the synaptic weights are updated according to Equation (9). This process is repeated until the stop condition is reached, which can be achieved when the error metric is lower than a target value or when the maximum number of epochs is reached. The Experimental data used as inputs and targets were normalised in the range of values between −1 and 1 (Equation (10)), as suggested in [64]:

$$x_n = -1 + 2 * \frac{x - x_{min}}{x_{max} - x_{min}}. \tag{10}$$

This type of normalisation limits the value of the data within the domain of the hyperbolic tangent function. In Equation (10), x represents either the input or output variables, x_n is the corresponding normalised value, x_{min} is the corresponding minimum value and x_{max} is the corresponding maximum value. The cross-validation technique was used to avoid the overfitting of the experimental data, which can lead to poor generalisation capability. The initial dataset of the experimental data is divided into three different sub-sets: the training, validation and test sets. Only the first of these is used to modify the parameters of the ANN, i.e., the synaptic weights. The others are needed to evaluate the performance of the trained neural network when it observes data that are not included in the training dataset. Hence, the partition percentages must be defined to perform the cross-validation technique. In this procedure, the following values were fixed:

- 60% for the training set;
- 20% for the validation set;
- 20% for the test set.

Random extractions are performed from the entire dataset to build these different clusters. Furthermore, the order of the examples within the same subset is randomised at the beginning of each epoch. The training algorithm stops when the error function related to the validation set reaches the desired value. The most common error function used within the EBP algorithm is the mean square error (*MSE*), which is calculated as follows:

$$MSE = \frac{\sum_{j=1}^{N_o} \sum_{m=1}^{p} \left(Y_{j,m} - y_{j,m}\right)^2}{p * N_o}. \tag{11}$$

In Equation (11), $Y_{j,m}$ is the target value of the j-th output for the m-th example, $y_{j,m}$ is the output value predicted by the ANN of the j-th output for the m-th example, and N_o is the number of output units. The latter assumes a value equal to 2 in this case. Furthermore, the mean absolute percentage error (*MAPE*), mean absolute error (*MAE*) and the determination coefficient (R^2) are evaluated as performance indexes. These error metrics are calculated, respectively, as follows:

$$MAPE_j = \left| \frac{\sum_{m=1}^{p} \frac{(Y_{j,m} - y_{j,m})}{Y_{j,m}}}{p} \right| * 100, \tag{12}$$

$$MAE_j = \left| \frac{\sum_{m=1}^{p} \left(Y_{j,m} - y_{j,m}\right)}{p} \right|, \tag{13}$$

$$R^2 = 1 - \frac{\sum_{j=1}^{N_o} \sum_{m=1}^{p} \left(Y_{j,m} - y_{j,m}\right)^2}{\sum_{j=1}^{N_o} \sum_{m=1}^{p} \left(Y_{j,m}\right)^2}. \tag{14}$$

The second step of the procedure, starting from the experimental data of the isothermal magnetisation and the calorimetric measurements obtained for different magnetic fields and absolute temperatures, provides an ANN-based analytical formulation of the magnetisation and specific heat of the investigated sample. Hence, bearing in mind Equation (3), making appropriate substitutions also considering Equation (10), these properties can be expressed as follows:

$$M(H,T) = M_{min} + \frac{M_{max} - M_{min}}{2} * \left(1 + \sum_{k=1}^{N_h} (w_{1k} * \tanh(v_h(H,T))) + b_{j=1}\right), \tag{15}$$

$$c_{P,H}(H,T) = c_{P,H_{min}} + \frac{c_{P,H_{max}} - c_{P,H_{min}}}{2} * \left(1 + \sum_{k=1}^{N_h} (w_{2k} * \tanh(v_h(H,T))) + b_{j=2}\right), \tag{16}$$

where:

$$v_h(H,T) = w_{k1} * \frac{H_{min} + H_{max} - 2H}{H_{min} - H_{max}} + w_{k2} * \frac{T_{min} + T_{max} - 2T}{T_{min} - T_{max}} + b_k. \tag{17}$$

The Equations (15) and (16) make it possible to obtain the characteristic curves of the magnetisation and specific heat as functions of the magnetic field and absolute temperature, for each value within the training domain of the ANN. In detail, Equation (15) is proposed as an alternative mathematical formulation of magnetisation that can be evaluated by different magnetic phenomenological models, most of them based on the Weiss Mean Field Theory (MFT). The Equations (15) and (16) depend on the parameters of the ANN, which are the synaptic weights and the minimum and maximum values identified during the normalisation process. The synaptic weights are grouped within the matrixes of the synaptic weights $W_h^{N_h \times (N_i+1)}$ and $W_o^{N_o \times (N_h+1)}$, organised as follows:

$$W_h = \begin{bmatrix} w_{11} & w_{12} & b_1 \\ w_{21} & w_{22} & b_2 \\ \vdots & \vdots & \vdots \\ w_{k1} & w_{k2} & b_k \end{bmatrix}, \tag{18}$$

$$W_o = \begin{bmatrix} w_{11} & w_{12} & \cdots & w_{1k} & b_1 \\ w_{21} & w_{22} & \cdots & w_{2k} & b_2 \end{bmatrix}, \tag{19}$$

where the subscript k identifies the k-th hidden neurons. Considering W_h, the first and the second column are referred to as the first and the second input, i.e., the applied magnetic field and the absolute temperature, respectively. In W_o, the first and the second row are referred to the first and the second output, i.e., the magnetisation and the specific heat, respectively. The minimum and the maximum value of the input and output variables are grouped into two matrixes, named $map_i^{2 \times N_i}$ and $map_t^{2 \times N_o}$, respectively. They are organised as follows:

$$map_i = \begin{bmatrix} H_{min} & T_{min} \\ H_{max} & T_{max} \end{bmatrix}, \tag{20}$$

$$map_t = \begin{bmatrix} M_{min} & c_{P,H_{min}} \\ M_{max} & c_{P,H_{max}} \end{bmatrix}. \tag{21}$$

The matrixes from Equations (18) and (21) represent the result of the second and the input for the third step of the procedure introduced here. It is important to highlight that the training dataset does not include all the available experimental data. By exploiting the generalisation capability of the ANN, a reduced number of experimental tests was needed to carry out the predictions for the different materials. Specifically, a sensitivity analysis considering the different sizes of the training set was performed to point out the proper dimension of the dataset. Seven different training sets were developed, changing the number of magnetic field samples from 3 to 101 and considering only 11 temperature values. In Figure 2 the results of this analysis are reported.

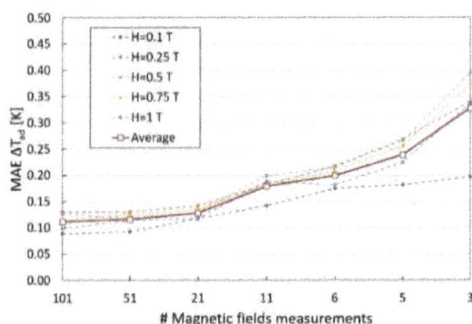

Figure 2. Mean absolute error of the adiabatic temperature change predictions with different sizes of the training set at different magnetic fields. The number of temperature samples considered during the training of the ANNs has been fixed at 11.

Hence, 21 different values of the magnetic field, from 0 T to 1 T with a step of 0.05 T, and 11 different values of the absolute temperature, from 270 K to 310 K with a step of 5 K, plus 250 K and 260 K, were used to perform the learning phase.

2.3. Isothermal Entropy-Change Evaluation

Using Equation (15), the isothermal entropy change of the investigated MCM can be obtained by numerical integration for every desired step of both the magnetic field and the temperature. The latter can lead to an improvement of the modelling capability of the material properties, since it is possible to compute the evaluation with a small temperature step, reducing the systematic errors [65]. However, the properties defined for the ANN developed in the previous step can be used to evaluate the isothermal entropy change of the MCM via an indirect method using an analytical approach (Equation (22)). The EBP algorithm requires continuous and differentiable functions to be performed. Hence, the magnetisation formulation of Equation (15) makes it possible to calculate the magnetisation derivative concerning absolute temperature at a constant magnetic field, as follows:

$$\left(\frac{\partial M(H,T)}{\partial T}\right)_H = \frac{M_{max} - M_{min}}{2} * \sum_{k=1}^{N_h}\left(\frac{2 * w_{1k} * w_{k2} * sech^2(v_h(H,T))}{T_{max} - T_{min}}\right). \tag{22}$$

In Equation (22), w_{1k} is the synaptic weight that links the first output, i.e., the magnetization of the specimen, to the k-th hidden neuron, whereas w_{k2} is the synaptic weight that links the k-th hidden neuron to the second input of the ANN, which is the absolute temperature T. It is important to note that the synaptic weights used in this equation are linked to the derivative argument M and the derivative variable T. The magnetization derivative value is used to calculate the isothermal entropy change using Maxwell's relation. Hence, by analytical integration of Equation (22), the isothermal entropy change can be expressed as:

$$\Delta s_{iso}(H_1, H_2, T) = \mu_0 \frac{M_{max} - M_{min}}{2} * \sum_{k=1}^{N_h}(B * C * (\tanh(v_h(H_2, T)) - \tanh(v_h(H_1, T)))), \tag{23}$$

where:

$$B = \frac{H_{max} - H_{min}}{T_{max} - T_{min}}, \tag{24}$$

$$C = \frac{w_{1k} * w_{k2}}{w_{k1}}. \tag{25}$$

Equation (23) represents a new mathematical formulation of the isothermal entropy change based on ANN theory. The values of H_2 and H_1 are the final and the initial external magnetic fields of the process to which the specimen is subjected, respectively. In Equation (25), w_{k1} is the synaptic weight that links the k-th hidden neuron to the first input of the ANN, which is the applied magnetic field H, i.e., the integration variable. Hence, using the output of the previous step, the isothermal entropy change can be straightforwardly obtained by Equation (23), avoiding the systematic errors caused by the numerical integration of Maxwell's relation.

2.4. Adiabatic Temperature-Change Evaluation

The adiabatic temperature change ΔT_{ad} can be obtained from the isothermal entropy change, but the most accurate method used in numerical modelling is based on the construction of the s-T diagram of the MCM. The latter ensures an accurate and coherent evaluation of the magnetocaloric properties of the materials, which are correlated with the thermodynamic relations and are strongly dependent on the temperature and the magnetic field. The success of an AMR numerical model is strongly related to the correct prediction of these properties. The s-T diagram is built using the calculation of the total entropy s_{tot} of the material. The evaluation of this property can be performed using different methods [37,40,66]. Another method is proposed in [67] where a protocol to perform the correct building of the s-T diagram for FOMT materials is described. In this procedure, the approach based on the use of the magnetisation data and the specific heat at zero magnetic field was implemented. The total entropy at zero field $s_{tot,H_0}(T)$ can be evaluated according to the following equation:

$$s_{tot,H_0}(T_1) = s_{ref} + \int_{T_{ref}}^{T_1} \frac{c_{P,H}(0,T)}{T} dT, \tag{26}$$

where s_{ref} and T_{ref} are the total entropy and the absolute temperature at the reference state, respectively, T_1 is the upper limit of the integration and $c_{P,H}(0,T)$ is the specific heat at zero magnetic field. For building the total entropy curves at different values of the magnetic field $s_{tot,H}$, it can proceed to add to isothermal entropy change calculated in the previous procedure step to the total entropy at zero magnetic field, as follows:

$$s_{tot,H}(H_{set},T) = s_{tot,H_0}(T) + \Delta s_{iso}(H_{set},T). \tag{27}$$

In Equation (27), H_{set} is the applied magnetic field to which the MCM is subjected. In this way, the s-T diagram is completed, and the adiabatic temperature change ΔT_{ad} can be computed according to Equation (28), where $T_2(s_{tot,H_2})$ is the temperature at the total entropy value along the curve of the magnetic field equal to H_2 and $T_1(s_{tot,H_1})$ is the temperature at the total entropy value along the curve of the magnetic field equal to H_1:

$$\Delta T_{ad} = T_2(s_{tot,H_2}) - T_1(s_{tot,H_1}). \tag{28}$$

Hence, the last step of the procedure described in this paper considers the isothermal entropy change as the input and provides the adiabatic temperature change as the output. The implementation of the entire process was made by developing a code written in the MATLAB environment, which allows loading of the experimental data, developing and training the ANN, and calculating the isothermal entropy change and the adiabatic temperature change. The specific heat values are directly provided by the ANN trained with both the magnetisation and heat-capacity measurements, although they should be calculated from the total entropy curves (see Equation (29)) to ensure the thermodynamic consistency of the data within the AMR numerical model:

$$c_{P,H}(T) = T * \left(\frac{\partial s_{tot,H}}{\partial T}\right)_{H=const}. \tag{29}$$

However, the evaluation of the magnetocaloric properties of the MCMs performed here can be carried out using only the parameters of the ANN, i.e., the synaptic weights and the input-output mapping. Hence, once the ANN is trained, it needs only a small database to store information about these parameters for different MCMs. The code can be easily integrated into the existing and new numerical models, like that recently introduced in Mugica et al. [68].

3. Results and Discussion

3.1. Procedure Performance with Gadolinium

The ANN (refer to Appendix A for the ANN parameters) fits well both in terms of the magnetisation curves and the specific heat behaviour, as well as with data that the ANN was not trained for. The model predicts very well the trend of the specific heat near the Curie temperature: the peak value of specific heat decreases as the magnetic field increases and it shifts towards higher temperatures (see Figure 3b). Considering both outputs of the ANN (magnetisation and specific heat), an average mean absolute percentage error (*MAPE*) equal to 7.0% and an average determination coefficient (R^2) of 0.9969 have been obtained.

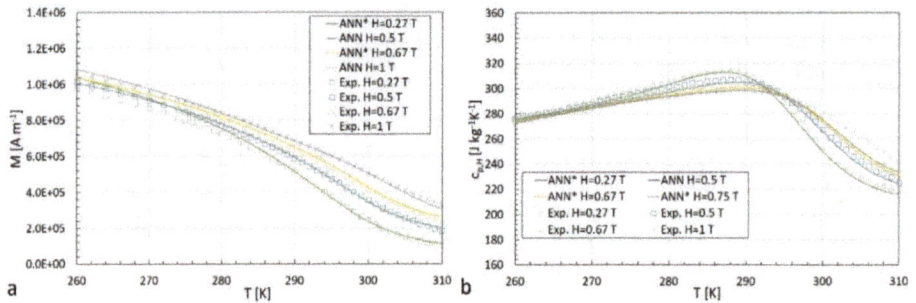

Figure 3. Measured and predicted value at different magnetic fields: (**a**) Magnetization of Gadolinium, (**b**) Specific heat at a constant magnetic field of Gadolinium. The asterisk in the legend indicates an applied magnetic field not included in the training dataset.

By performing the third step of the procedure, the isothermal entropy change was calculated for the same conditions (magnetic fields and absolute temperatures) as in the previous step (see Figure 4). It is highlighted that in most cases the model predicts values of isothermal entropy change within the range of the error of the numerical solution [37].

The trend of isothermal entropy change as a function of temperature is also well predicted around the Curie temperature, where the isothermal entropy change shows its peak value.

From Figure 5b it is clear that the application of the ANN-based procedure leads to a slight underestimation of the adiabatic temperature change for all the magnetic fields, except for the maximum one. However, although these deviations are more pronounced for smaller magnetic fields, these differences are within the range of the error. Generally, it emerges that the trend of the adiabatic temperature changes as a function of temperature is preserved, also around the Curie temperature. It can be concluded that the model provides an estimation of the ΔT_{ad} with a mean absolute error (*MAE*) equal to 0.1 *K* and a maximum absolute error of 0.4 *K*. The determination coefficient between the measured and predicted values of the adiabatic temperature change is equal to 0.9871, which is close to the value obtained by Mugica Guerrero et al. [48].

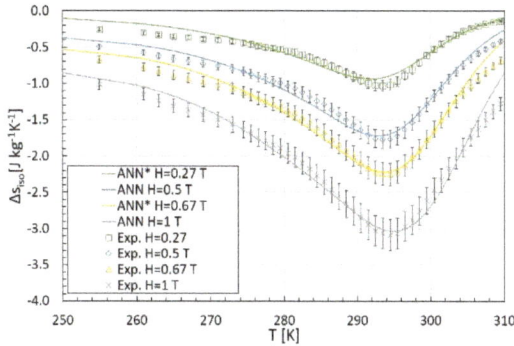

Figure 4. Comparison between the isothermal entropy changes of gadolinium at different magnetic fields calculated with the ANN approach and those calculated by direct numerical integration. The asterisk in the legend indicates an applied magnetic field not included in the training dataset.

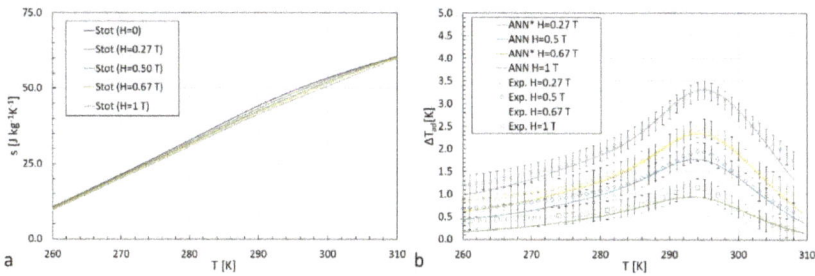

Figure 5. Results of the last step of the procedure: (**a**) *s-T* diagram of Gd and (**b**) adiabatic temperature change of gadolinium during the magnetisation phase at different magnetic fields. The asterisk in the legend indicates an applied magnetic field not included in the training dataset.

In Figure 6 a comparison between the ANN-based procedure (introduced in this study), the direct-measurement method, and the MFT reported in Petersen et al. [45] is shown. It is evident that the MFT does not fit the experimental values of the specific heat particularly well. The values predicted by the ANN-based procedure are much closer to the measured ones, especially for $H = 1\ T$. At zero field, some differences in the specific heat values occur around the Curie temperature. These deviations could be reduced by considering a smaller temperature step in the training set of the ANN around the Curie temperature. Furthermore, the MFT overestimates the adiabatic temperature change below the Curie temperature, with the most significant deviation from the experimental data at the Curie temperature. For temperatures above the Curie temperature, the deviations decrease. On the other hand, the prediction of the adiabatic temperature change using the developed ANN-based procedure is significantly better, within the error range of the experimental data (see Figure 6b), excluding just a few points at higher temperatures (between 306 K and 308 K).

3.2. Procedure Performance with the La-Fe-Co-Si Alloy

As shown in Figure 7, the predicted magnetisation and the specific heat values for the three specimens of La-Fe-Co-Si alloy fit well with the experimental results. For the Specimens 1–3, the *MAPE* values are equal to 8.5%, 6.6% and 6.5%, whereas the determination coefficients are 0.9980, 0.9946 and 0.9943, respectively.

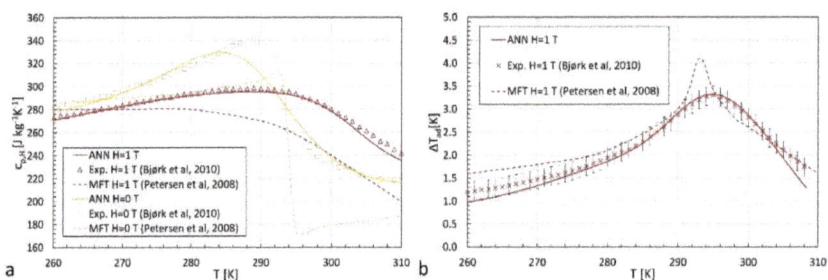

Figure 6. Comparison with MFT approach: (**a**) Measured values of specific heat of Gd at $H = 1\,T$ and $H = 0\,T$ in comparison with values predicted by the MFT approach and ANN-based procedure, (**b**) Measured values of the adiabatic temperature change of Gd at $H = 1\,T$ in comparison with the values predicted by the MFT approach and the ANN-based procedure.

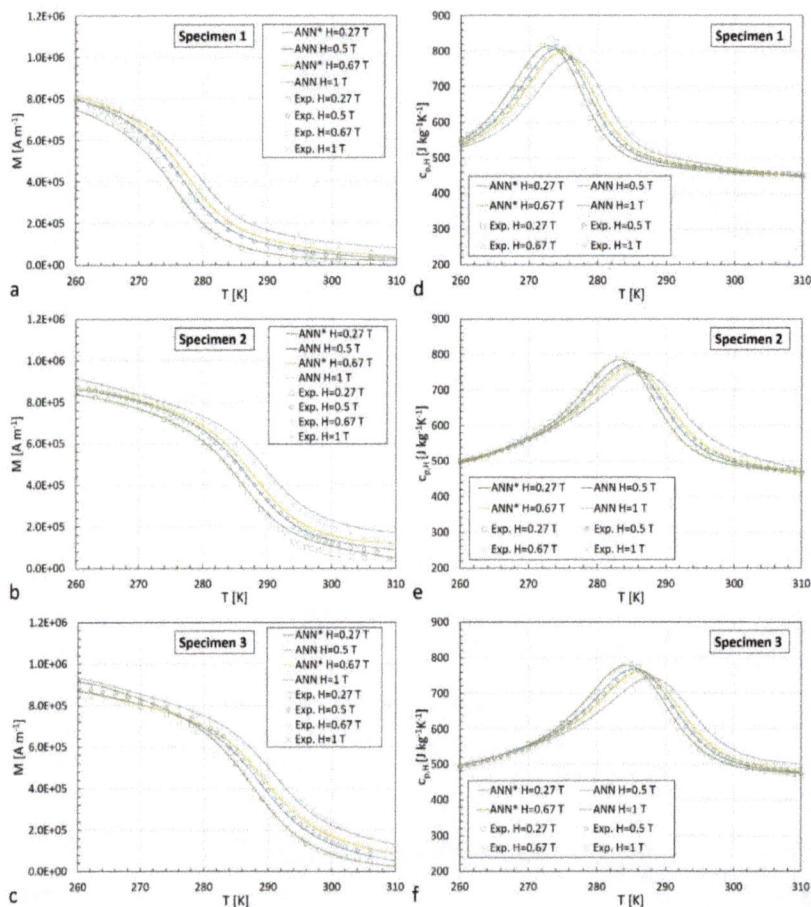

Figure 7. Measured and predicted values at different magnetic fields: (**a–c**) Magnetization of the three samples of La-Fe-Co-Si alloy, (**d–f**) Specific heat at a constant magnetic field of the three samples of La-Fe-Co-Si alloy. The asterisk in the legend indicates an applied magnetic field not included in the training dataset.

The comparison of the isothermal entropy changes calculated with the ANN-based procedure and those obtained by the direct numerical integration of Maxwell's relation shows a good agreement (see Figure 8). It is worth pointing out that, also in these cases, the ANN-based calculation process maintains the behaviour of the isothermal entropy change, highlighting a peak value around the Curie temperature for all the specimens. However, the predicted values for the isothermal entropy change of the three specimens are within the range of the error [37] in most cases.

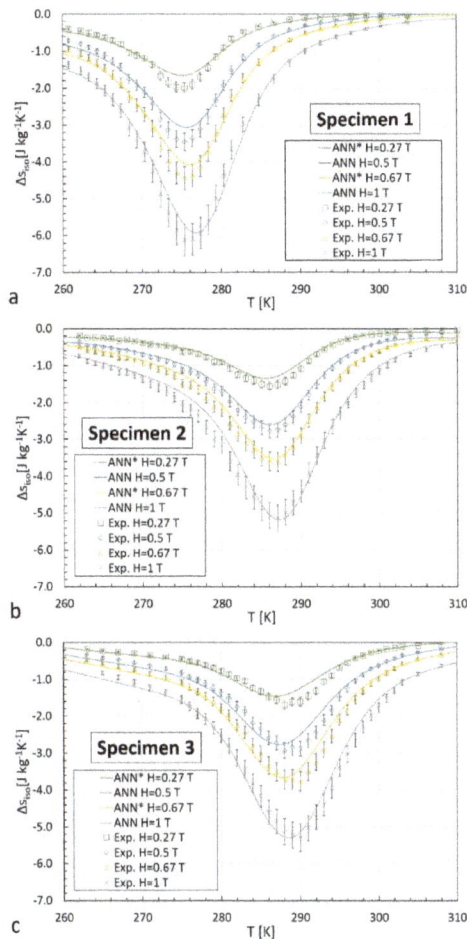

Figure 8. Comparison between the isothermal entropy changes of the three samples of La-Fe-Co-Si alloy at different magnetic fields calculated with the ANN approach and those calculated by numerical integration. The asterisk in the legend indicates an applied magnetic field not included in the training dataset.

The outcomes calculated in the fourth step of the procedure for the three samples of La-Fe-Co-Si alloy are shown in Figure 9, where in the left column (a–c) are the *s-T* diagrams, while in the right one (d–f), the adiabatic temperature changes are shown for four applied magnetic field values.

For all three samples, the *MAE* value related to the adiabatic temperature change was kept at 0.1 *K*, with a maximum between 0.2 *K* and 0.3 *K*. On the other hand, the R^2 values for Specimens 1–3 are equal to 0.9932, 0.9845 and 0.9907, respectively. The parameters of the ANNs developed for the La-Fe-Co-Si alloys are reported in Appendix B.

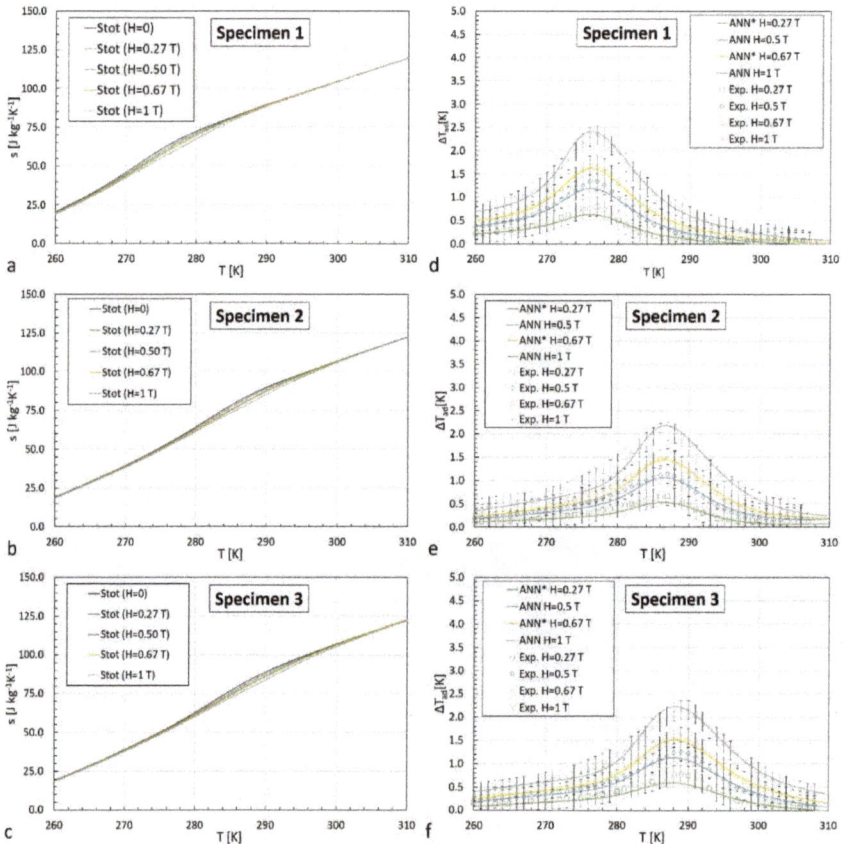

Figure 9. Results of the last step of the procedure: (**a–c**) *s-T* diagram of the three samples of La-Fe-Co-Si alloy and (**d–f**) adiabatic temperature change of the three samples of La-Fe-Co-Si alloy during the magnetisation phase at different magnetic fields. The asterisk in the legend indicates an applied magnetic field not included in the training dataset.

3.3. Summary of the Results

In Table 1 the maximum values of the adiabatic temperature change, the isothermal entropy change and the specific heat of the investigated MCMs are reported. These results show good agreement with the experimental data reported in Bjørk et al. [26], considering both the maximum values of the properties and the temperature at which these maxima occur (T_{peak}).

The performances shown by the ANN-based procedure with the different MCMs are summarised in Table 2 with respect to the error metrics (Section 2). In the second column the number of hidden neurons (N_h) is reported.

Considering the accuracy of the temperature sensor, the results reported in the previous sections and summarised in Table 2 prove the excellent prediction capability of the method. Hence, the application of the ANN-based procedure for two different materials led to similar error values ($R^2 = 0.9871$ for Gd against an average $R^2 = 0.9895$ for La-Fe-Co-Si alloy), demonstrating its capability to model the magnetocaloric properties of different types of MCMs.

Table 1. Maximum values of the magnetocaloric properties of Gd and the three samples of La-Fe-Co-Si alloy at $\mu 0H = 1\,T$ from Bjørk et al. [26] and calculated by the ANN-based procedure.

	Gadolinium		Specimen 1		Specimen 2		Specimen 3	
	Bjørk et al. [26]	This work	Bjørk et al. [26]	This work	Bjørk et al. [26]	This work	Bjørk et al. [26]	This work
ΔT_{ad} (1 T) [K]	3.3	3.3 (0.0%)	2.3	2.4 (+4.3%)	2.1	2.2 (+4.8%)	2.1	2.2 (+4.8%)
T_{peak} [K]	295.1	294.9	277.1	276.3	287.1	286.7	289.6	288.1
$|\Delta s_{iso}$ (1 T)| [J/kgK]	3.1	3.0 (−3.2%)	6.2	5.9 (−4.8%)	5.1	5.2 (+2.0%)	5.0	5.3 (+6.0%)
T_{peak} [K]	294.8	295.0	275.8	277.0	287.1	287.0	289.8	289.0
$c_{p,H}$ (1 T) [J/kgK]	298.8	295.0 (−1.3%)	783.4	776.6 (−0.9%)	754.9	748.8 (−0.8%)	740.9	743.7 (+0.4%)
T_{peak} [K]	289.2	289.0	276.1	276.0	286.1	286.0	288.2	288.0

Table 2. Summary of the results carried out by implementing the ANN-based procedure with Gd and the three samples of La-Fe-Co-Si alloy.

MCM	N_h	MAE M [A/m]	MAE $c_{p,H}$ [J/kgK]	MAPE M [%]	MAPE $c_{p,H}$ [%]	MAPE Ave [%]	R^2	R^2	ΔT_{ad} MAE [K]	ΔT_{ad} E_{max} [K]
Gd	15	20681.1	2.6	13.0%	0.9%	7.0%	0.9969	0.9871	0.12	0.4
LaFe$_{11.06}$Co$_{0.86}$Si$_{1.08}$	13	13504.0	7.4	15.8%	1.2%	8.5%	0.9980	0.9932	0.04	0.2
LaFe$_{11.05}$Co$_{0.94}$Si$_{1.01}$	13	22975.9	3.5	12.7%	0.6%	6.6%	0.9946	0.9845	0.06	0.3
LaFe$_{10.96}$Co$_{0.97}$Si$_{1.07}$	14	21518.0	5.3	12.1%	0.9%	6.5%	0.9943	0.9907	0.05	0.2

The method introduced in this work provides thermodynamic properties for both evaluated groups of MCMs, making it possible to predict the magnetocaloric properties of both with reasonable accuracy. Furthermore, few data were used to train the ANN model, as described in Section 2. This fact represents a significant advantage, since the time needed to carry out the experiments decreases significantly. Indeed, computing the numerical integration with the same amount of data as used by the ANN-procedure could lead to up to 35% deviations of the maximum isothermal entropy change compared to the full data set measured with a step of 1 K [65].

Furthermore, the training parameters were kept unchanged. This makes it possible to perform the ANN procedure efficiently with different MCMs, by facilitating and speeding up the characterisation of new materials. To evaluate the magnetocaloric properties of an MCM by this procedure, one should measure the magnetisation and the specific heat at different temperatures and magnetic fields. Considering an arbitrary range, the temperature and the magnetic field steps which can be considered for the experiments are 5 K and 0.1 T, respectively. These values represent a guideline stem from the results reported in this work (see Figure 2). The volume of collected data depends on the chosen range. For example, if one wants to model the behaviour of an MCM in the magnetic field range between 0 T and 0.5 T (6 different values) and in the temperature range between 250 K and 350 K (21 different values), the volume of collected data will be equal to 126 for each output of the ANN, i.e., magnetisation and specific heat. Once the ANN is trained, the results could be verified and validated by analysing some intermediate points not considered during the ANN training. Hence, just a few other experiments are needed to validate the new ANN. However, some precautions have been taken when using the ANN-based procedure, related to the experimental part of the process. The results are strongly affected by the training phase of the ANN since the main thermodynamic properties are calculated using the synaptic weights identified during this step. The quality of the learning depends on the quality of the data with which the ANN is trained. Hence, it is necessary to make correct measurements of both the specific heat and the magnetisation. For the latter case, some indications can be found in the literature [41,65,69].

Energies **2019**, *12*, 1871

4. Conclusions

We propose a new and flexible procedure based on the use of ANNs to evaluate the thermodynamic properties of MCMs. It is shown that the procedure improves over the commonly used methods in the framework of magnetic refrigeration and provides similar or better results based on a reduced amount of information. Indeed, only a few experimental data are needed to perform a complete thermodynamic characterisation of an MCM, and no modifications to its general formulation are required. Hence, the tool developed in this work, available from the authors upon reasonable request, is proposed as a standardised procedure to evaluate the magnetocaloric properties of MCMs, which can improve the implementation of AMR numerical models and speed up the characterisation of new MCMs.

Author Contributions: A.M. and J.T. conceived the idea and helped with the procedure development. M.G.D.D. developed the procedure code and wrote the paper. U.T. managed the experimental data and helped with the procedure development. A.K. and C.A. supervised the entire work.

Funding: This research received no external funding.

Acknowledgments: The authors would like to thank Rasmus Bjørk from Technical University of Denmark for kindly sharing with us measured values of magnetocaloric properties for Gd and La-Fe-Co-Si alloys.

Conflicts of Interest: The authors declare no conflict of interest.

Appendix A

Table A1. Hidden synaptic weights of Gd ANN.

W_h	H	T	b
w_1	−1.4212	−0.2781	−1.5975
w_2	−1.7606	−0.1633	−1.8839
w_3	0.0286	0.9740	0.0373
w_4	2.2650	0.3972	1.4500
w_5	−1.1547	0.4693	−0.8224
w_6	0.9427	−2.6658	1.2563
w_7	−0.2284	0.2009	−0.6378
w_8	0.8338	0.2446	0.6600
w_9	0.1867	−0.5571	−0.7011
w_{10}	−0.6442	3.2147	−1.9418
w_{11}	0.2429	−0.2356	1.0931
w_{12}	0.0053	−1.7567	0.2993
w_{13}	−0.8154	0.3201	−0.2907
w_{14}	−0.1940	0.6915	−0.2957
w_{15}	0.0960	−0.7824	−0.1980

Table A2. Output synaptic weights of Gd ANN.

W_o^T	M	$c_{p,H}$
w_1	−1.5651	−0.6911
w_2	−1.5142	0.6707
w_3	−0.2390	0.5765
w_4	−1.2677	−0.0193
w_5	−0.4934	−0.3781
w_6	0.0465	−1.1169
w_7	0.5484	−0.1520
w_8	0.1967	−0.3960
w_9	0.8164	0.0993
w_{10}	−0.2668	−1.7536
w_{11}	−1.2219	0.4227
w_{12}	0.2918	0.4122
w_{13}	0.1137	0.0266
w_{14}	0.2462	0.0562
w_{15}	−0.0229	−0.6135
b	−0.3358	−1.0183

Table A3. Input-target mapping of Gd ANN.

	H [T]	T [K]	M [A/m]	$c_{p,H}$ [J/kgK]
Min	0	250	0	216.1
Max	1	310	1144982	339.0

Appendix B

Table A4. Hidden synaptic weights of La-Fe-Co-Si ANNs.

	Specimen 1			Specimen 2			Specimen 3		
W_h	H	T	b	H	T	b	H	T	b
w_1	0.1917	0.8973	0.7315	0.4618	−1.6667	0.3750	0.8520	0.2031	0.8984
w_2	−0.2097	0.3061	−1.2611	−0.3502	4.7871	−1.1755	−0.1914	0.5909	−1.1288
w_3	−0.3497	−0.2112	0.0701	1.3672	−0.3242	1.1277	0.1162	−0.1996	−0.2653
w_4	−0.4082	5.2952	0.7372	−0.1936	1.7104	0.6309	−0.4516	0.3241	−0.0616
w_5	0.4184	−1.4710	0.0050	−0.4580	3.9225	−0.4588	0.4635	0.6240	0.3144
w_6	−0.3845	3.7749	1.0538	−0.0357	0.0634	0.9233	−0.3054	4.0817	−0.9671
w_7	−0.4790	0.8194	−0.7946	−0.5250	−0.6962	−0.5996	−0.8399	0.9857	−0.5522
w_8	−0.0558	−0.8351	0.4745	−0.0765	−0.6675	0.0295	−0.0363	1.3581	0.4379
w_9	0.5199	−0.5261	0.4265	0.7811	1.0029	−0.0288	1.8243	0.4333	2.2864
w_{10}	−0.0324	−0.0243	0.7000	0.3465	−0.5162	−0.5920	−0.4274	4.0521	−0.3688
w_{11}	−0.5055	0.7145	0.5963	−0.0238	0.9699	−0.4976	−0.1458	0.4901	0.3521
w_{12}	−0.3771	0.2812	−0.2925	−1.6583	−0.2331	−0.7204	1.8930	0.3430	1.2621
w_{13}	−0.7053	−0.4690	0.4303	−0.7434	−0.6370	0.6484	−1.1674	0.6076	0.0530
w_{14}							−0.1267	0.1076	−1.4187

Table A5. Output synaptic weights of La-Fe-Co-Si ANNs.

	Specimen 1		Specimen 2		Specimen 3	
W_o^T	M	$c_{p,H}$	M	$c_{p,H}$	M	$c_{p,H}$
w_1	−0.1875	0.2929	0.6240	0.6533	0.7351	−0.6447
w_2	0.4129	0.4394	−0.3269	−2.3121	0.7421	−0.3485
w_3	−0.0026	−0.7827	1.2805	−0.0905	0.1774	0.5389
w_4	−0.3346	−2.3002	−0.1154	0.8733	−0.2926	0.1307
w_5	0.3067	0.5734	−0.0184	2.2231	0.2230	0.2111
w_6	−0.1666	2.3449	−0.3507	−0.8183	−0.4881	−2.3771
w_7	0.1253	−0.3257	−1.0127	0.2958	−0.7768	−0.1102
w_8	−0.7623	0.2733	−0.5434	−0.6073	−0.2466	0.7265
w_9	0.2552	−0.3323	−0.4603	0.2283	1.9406	0.0475
w_{10}	0.2185	−0.4438	1.0259	0.3566	0.0306	1.9483
w_{11}	−0.1077	0.2698	0.8814	−0.3649	−0.6731	0.0233
w_{12}	−0.9737	0.4039	0.8366	−0.0055	−0.8551	0.1545
w_{13}	0.6014	0.1432	0.4854	0.2014	0.6048	−0.0711
w_{14}					1.2991	0.4230
w_{15}						
b	0.0585	−0.4769	−0.1803	−0.3426	−0.0325	−0.5758

Table A6. Input-target mapping of La-Fe-Co-Si ANNs.

		H [T]	T [K]	M [A/m]	$c_{p,H}$ [J/kgK]
Specimen 1	Min	0	252	0	450
	Max	1	311	879830	831
Specimen 2	Min	0	252	0	468
	Max	1	309	944056	792
Specimen 3	Min	0	252	0	466
	Max	1	311	945918	826

References

1. Qian, S.; Nasuta, D.; Rhoads, A.; Wang, Y.; Geng, Y.; Hwang, Y.; Radermacher, R.; Takeuchi, I. Not-in-kind cooling technologies: A quantitative comparison of refrigerants and system performance. *Int. J. Refrig.* **2016**, *62*, 177–192. [CrossRef]

2. Bansal, P.; Vineyard, E.; Abdelaziz, O. Status of not-in-kind refrigeration technologies for household space conditioning, water heating and food refrigeration. *Int. J. Sustain. Built Environ.* **2012**, *1*, 85–101. [CrossRef]

3. Kitanovski, A.; Plaznik, U.; Tomc, U.; Poredoš, A. Present and future caloric refrigeration and heat-pump technologies. *Int. J. Refrig.* **2015**, *57*, 288–298. [CrossRef]

4. US Department of Energy—Office of Energy Efficiency and Renewable Energy. Using Magnets to Keep Cool: Breakthrough Technology Boosts Energy Efficiency of Refrigerators. 2014. Available online: https://www.energy.gov/eere/articles/using-magnets (accessed on 23 July 2018).

5. US Department of Energy—Office of Energy Efficiency and Renewable Energy. ORNL Refrigerator Cools with Magnetism, Not Freon. 2016. Available online: https://www.energy.gov/eere/buildings/articles/orn (accessed on 23 July 2018).

6. EU Project 603885 Final Report Summary—ELICIT (Environmentally Low Impact Cooling Technology). 2017. Available online: https://cordis.europa.eu/result/rcn/201566_en.html (accessed on 25 July 2018).

7. EU Project 214864 Final Report Summary—SSEEC (Solid State Energy Efficient Cooling). 2013. Available online: https://cordis.europa.eu/result/rcn/57173_en.html (accessed on 25 July 2018).

8. UNEP. Twenty-Eighth Meeting of the Parties to the Montreal Protocol on Substances that Deplete the Ozone Layer. 2016. 1-9. Decision XXVIII/— Further Amendment of the Montreal Protocol.

9. The European Commission. Regulation (EU) No 517/2014 of the European Parliament and of the Council of 16 April 2014 on Fluorinated Greenhouse Gases and Repealing Regulation (EC) No 842/2006. 2014. Available online: https://www.eea.europa.eu/policy-documents/regulation-eu-no-517-2014 (accessed on 26 July 2018).

10. McLinden, M.O.; Brown, J.S.; Brignoli, R.; Kazakov, A.F.; Domanski, P.A. Limited options for low-global-warming-potential refrigerants. *Nat. Commun.* **2017**, *8*, 1–9. [CrossRef] [PubMed]

11. Bansal, P.; Vineyard, E.; Abdelaziz, O. Advances in household appliances—A review. *Appl. Therm. Eng.* **2011**, *31*, 3748–3760. [CrossRef]

12. Aprea, C.; Greco, A.; Maiorino, A. The application of a desiccant wheel to increase the energetic performances of a transcritical cycle. *Energy Convers. Manag.* **2015**, *89*, 222–230. [CrossRef]

13. Llopis, R.; Cabello, R.; Sánchez, D.; Torrella, E. Energy improvements of CO_2 transcritical refrigeration cycles using dedicated mechanical subcooling. *Int. J. Refrig.* **2015**, *55*, 129–141. [CrossRef]

14. Weiss, P.; Piccard, A. Le phénomène magnétocalorique. *J. Phys. Théorique Appliquée* **1917**, *7*, 103–109. [CrossRef]

15. Gschneidner, K.A.; Pecharsky, V.K. Rare earths and magnetic refrigeration. *J. Rare Earths* **2006**, *24*, 641–647. [CrossRef]

16. Zverev, V.I.; Tishin, A.M.; Kuz'Min, M.D. The maximum possible magnetocaloric ΔT effect. *J. Appl. Phys.* **2010**, *107*, 043907. [CrossRef]

17. Barclay, J.A. Use of a ferrofluid as the heat-exchange fluid in a magnetic refrigerator. *J. Appl. Phys.* **1982**, *53*, 2887–2894. [CrossRef]

18. Brown, G.V. Magnetic heat pumping near room temperature. *J. Appl. Phys.* **1976**, *47*, 3673–3680. [CrossRef]

19. Aprea, C.; Greco, A.; Maiorino, A.; Mastrullo, R.; Tura, A. Initial experimental results from a rotary permanent magnet magnetic refrigerator. *Int. J. Refrig.* **2014**, *43*, 111–122. [CrossRef]

20. Engelbrecht, K.; Eriksen, D.; Bahl, C.R.H.; Bjørk, R.; Geyti, J.; Lozano, J.A.; Nielsen, K.K.; Saxild, F.; Smith, A.; Pryds, N. Experimental results for a novel rotary active magnetic regenerator. *Int. J. Refrig.* **2012**, *35*, 1498–1505. [CrossRef]

21. Lozano, J.A.; Capovilla, M.S.; Trevizoli, P.V.; Engelbrecht, K.; Bahl, C.R.H.; Barbosa, J.R. Development of a novel rotary magnetic refrigerator. *Int. J. Refrig.* **2016**, *68*, 187–197. [CrossRef]

22. Romero Gómez, J.; Ferreiro Garcia, R.; Carbia Carril, J.; Romero Gómez, M. Experimental analysis of a reciprocating magnetic refrigeration prototype. *Int. J. Refrig.* **2013**, *36*, 1388–1398. [CrossRef]

23. Tagliafico, L.A.; Scarpa, F.; Valsuani, F.; Tagliafico, G. Preliminary experimental results from a linear reciprocating magnetic refrigerator prototype. *Appl. Therm. Eng.* **2013**, *52*, 492–497. [CrossRef]

24. Tušek, J.; Zupan, S.; Šarlah, A.; Prebil, I.; Poredoš, A. Development of a rotary magnetic refrigerator. *Int. J. Refrig.* **2010**, *33*, 294–300. [CrossRef]

25. Tura, A.; Rowe, A. Permanent magnet magnetic refrigerator design and experimental characterization. *Int. J. Refrig.* **2011**, *34*, 628–639. [CrossRef]

26. Bjørk, R.; Bahl, C.R.H.; Katter, M. Magnetocaloric properties of LaFe13-x-yCoxSi y and commercial grade Gd. *J. Magn. Magn. Mater.* **2010**, *322*, 3882–3888. [CrossRef]

27. Balli, M.; Sari, O.; Zamni, L.; Mahmed, C.; Forchelet, J. Implementation of La(Fe, Co)13-xSixmaterials in magnetic refrigerators: Practical aspects. *Mater. Sci. Eng. B Solid-State Mater. Adv. Technol.* **2012**, *177*, 629–634. [CrossRef]

28. Legait, U.; Guillou, F.; Kedous-Lebouc, A.; Hardy, V.; Almanza, M. An experimental comparison of four magnetocaloric regenerators using three different materials. *Int. J. Refrig.* **2014**, *37*, 147–155. [CrossRef]

29. Tušek, J.; Kitanovski, A.; Tomc, U.; Favero, C.; Poredoš, A. Experimental comparison of multi-layered La-Fe-Co-Si and single-layered Gd active magnetic regenerators for use in a room-temperature magnetic refrigerator. *Int. J. Refrig.* **2014**, *37*, 117–126. [CrossRef]

30. Pulko, B.; Tušek, J.; Moore, J.D.; Weise, B.; Skokov, K.; Mityashkin, O.; Kitanovski, A.; Favero, C.; Fajfar, P.; Gutfleisch, O.; et al. Epoxy-bonded La–Fe–Co–Si magnetocaloric plates. *J. Magn. Magn. Mater.* **2015**, *375*, 65–73. [CrossRef]

31. Kubacki, J.; Balin, M.K.; Kulpa, M.; Hawełek, Ł.; Włodarczyk, P.; Zackiewicz, P.; Kowalczyk, M.; Polak, M.; Szade, J. Magnetic moments and exchange splitting in Mn3s and Mn2p core levels of magnetocaloric Mn1.1Fe0.9P0.6As0.4and Mn1.1Fe0.9P0.5As0.4Si0.1compounds. *Phys. B Condens. Matter* **2017**, *2*, 2–7.

32. Szymczak, R.; Nedelko, N.; Lewińska, S.; Zubov, E.; Sivachenko, A.; Gribanov, I.; Radelytskyi, I.; Dyakonov, K.; Ślawska-Waniewska, A.; Valkov, V.; et al. Comparison of magnetocaloric properties of the Mn2-xFexP0.5As0.5(x = 1.0 and 0.7) compounds. *Solid State Sci.* **2014**, *36*, 29–34. [CrossRef]

33. Zhang, H.; Gimaev, R.; Kovalev, B.; Kamilov, K.; Zverev, V.; Tishin, A. Review on the materials and devices for magnetic refrigeration in the temperature range of nitrogen and hydrogen liquefaction. *Phys. B Condens. Matter* **2019**, *558*, 65–73. [CrossRef]

34. Gimaev, R.; Spichkin, Y.; Kovalev, B.; Kamilov, K.; Zverev, V.; Tishin, A. Review on magnetic refrigeration devices based on HTSC materials. *Int. J. Refrig.* **2019**, *100*, 1–12. [CrossRef]

35. Nielsen, K.K.; Tusek, J.; Engelbrecht, K.; Schopfer, S.; Kitanovski, A.; Bahl, C.R.H.; Smith, A.; Pryds, N.; Poredos, A. Review on numerical modeling of active magnetic regenerators for room temperature applications. *Int. J. Refrig.* **2011**, *34*, 603–616. [CrossRef]

36. Kamran, M.S.; Sun, J.; Tang, Y.B.; Chen, Y.G.; Wu, J.H.; Wang, H.S. Numerical investigation of room temperature magnetic refrigerator using microchannel regenerators. *Appl. Therm. Eng.* **2016**, *102*, 1126–1140. [CrossRef]

37. Pecharsky, V.K.; Gschneidner, K.A. Magnetocaloric effect from indirect measurements: Magnetization and heat capacity. *J. Appl. Phys.* **1999**, *86*, 565–575. [CrossRef]

38. Dan'kov, S.Y.; Tishin, A.; Pecharsky, V.; Gschneidner, K. Magnetic phase transitions and the magnetothermal properties of gadolinium. *Phys. Rev. B Condens. Matter Mater. Phys.* **1998**, *57*, 3478–3490. [CrossRef]

39. Lee, J.S. Evaluation of the magnetocaloric effect from magnetization and heat capacity data. *Phys. Status Solidi Basic Res.* **2004**, *241*, 1765–1768. [CrossRef]

40. Nielsen, K.K.; Bez, H.N.; Von Moos, L.; Bjørk, R.; Eriksen, D.; Bahl, C.R.H. Direct measurements of the magnetic entropy change. *Rev. Sci. Instrum.* **2015**, *86*, 103903. [CrossRef]

41. Franco, V.; Blázquez, J.S.; Ipus, J.J.; Law, J.Y.; Moreno-Ramírez, L.M.; Conde, A. Magnetocaloric effect: From materials research to refrigeration devices. *Prog. Mater. Sci.* **2018**, *93*, 112–232. [CrossRef]

42. Tušek, J.; Kitanovski, A.; Prebil, I.; Poredoš, A. Dynamic operation of an active magnetic regenerator (AMR): Numerical optimization of a packed-bed AMR. *Int. J. Refrig.* **2011**, *34*, 1507–1517. [CrossRef]

43. Liu, M.; Yu, B. Numerical investigations on internal temperature distribution and refrigeration performance of reciprocating active magnetic regenerator of room temperature magnetic refrigeration. *Int. J. Refrig.* **2011**, *34*, 617–627. [CrossRef]

44. Aprea, C.; Maiorino, A. A flexible numerical model to study an active magnetic refrigerator for near room temperature applications. *Appl. Energy* **2010**, *87*, 2690–2698. [CrossRef]

45. Petersen, T.F.; Pryds, N.; Smith, A.; Hattel, J.; Schmidt, H.; Høgaard Knudsen, H.J. Two-dimensional mathematical model of a reciprocating room-temperature Active Magnetic Regenerator. *Int. J. Refrig.* **2008**, *31*, 432–443. [CrossRef]

46. Brown, G.V. Magnetic Stirling Cycles—A new application for magnetic materials. *IEEE Trans. Magn.* **1977**, *13*, 1146–1148. [CrossRef]

47. Paudyal, D.; Pecharsky, V.K.; Gschneidner, K.A.; Harmon, B.N. Electron correlation effects on the magnetostructural transition and magnetocaloric effect in Gd5 Si2 Ge2. *Phys. Rev. B Condens. Matter Mater. Phys.* **2006**, *73*, 1–12. [CrossRef]

48. Mugica Guerrero, I.; Poncet, S.; Bouchard, J. Entropy generation in a parallel-plate active magnetic regenerator with insulator layers. *J. Appl. Phys.* **2017**, *121*, 074901. [CrossRef]

49. Chiba, Y.; Marif, Y.; Henini, N.; Tlemcani, A. Artificial Neural Networks Modeling of an Active Magnetic Refrigeration Cycle. In *Artificial Intelligence in Renewable Energetic Systems. ICAIRES 2017*; Lecture Notes in Networks and Systems; Hatti, M., Ed.; Springer: Cham, Switzerland, 2018; Volume 35.

50. Aprea, C.; Greco, A.; Maiorino, A. An application of the artificial neural network to optimise the energy performances of a magnetic refrigerator. *Int. J. Refrig.* **2017**, *82*, 238–251. [CrossRef]

51. Hartenstein, T.; Li, C.; Lefkidis, G.; Hübner, W. Local light-induced spin manipulation in two magnetic centre metallic chains. *J. Phys. D Appl. Phys.* **2008**, *41*, 164006. [CrossRef]

52. Neufcourt, L.; Cao, Y.; Nazarewicz, W.; Viens, F. Bayesian approach to model-based extrapolation of nuclear observables. *Phys. Rev. C* **2018**, *98*, 1–17. [CrossRef]

53. Pennington, R.S.; Coll, C.; Estradé, S.; Peiró, F.; Koch, C.T. Neural-network-based depth-resolved multiscale structural optimization using density functional theory and electron diffraction data. *Phys. Rev. B* **2018**, *97*, 1–10. [CrossRef]

54. Jeppesen, S.; Linderoth, S.; Pryds, N.; Kuhn, L.T.; Jensen, J.B. Indirect measurement of the magnetocaloric effect using a novel differential scanning calorimeter with magnetic field. *Rev. Sci. Instrum.* **2008**, *79*, 083901. [CrossRef]

55. Bianchini, M.; Scarselli, F. On the complexity of neural network classifiers: A comparison between shallow and deep architectures. *IEEE Trans. Neural Netw. Learn. Syst.* **2014**, *25*, 1553–1565. [CrossRef] [PubMed]

56. Cybenko, G. Correction: Approximation by superpositions of a sigmoidal function. *Math. Control. Signals Syst.* **1989**, *2*, 303–314. [CrossRef]

57. Sheela, K.G.; Deepa, S.N. Review on methods to fix number of hidden neurons in neural networks. *Math. Probl. Eng.* **2013**, *2013*, 425740. [CrossRef]

58. Mohanraj, M.; Jayaraj, S.; Muraleedharan, C. Applications of artificial neural networks for refrigeration, air-conditioning and heat pump systems—A review. *Renew. Sustain. Energy Rev.* **2012**, *16*, 1340–1358. [CrossRef]

59. Hunter, D.; Yu, H.; Pukish, M.S.; Kolbusz, J.; Wilamowski, B.M. Selection of Proper Neural Network Sizes and Architectures—A Comparative Study. *IEEE Trans. Ind. Inform.* **2012**, *8*, 228–240. [CrossRef]

60. Qian, G.; Yong, H. Forecasting the Rural Per Capita Living Consumption Based on Matlab BP Neural Shanghai University of Engineering Science. *Int. J. Bus. Soc. Sci.* **2013**, *4*, 131–137.

61. Rumelhart, D.E.; Hinton, G.E.; Williams, R.J. Learning representations by back-propagating errors. *Nature* **1986**, *323*, 533–536. [CrossRef]

62. Prechelt, L. Automatic early stopping using cross validation: Quantifying the criteria. *Neural Netw.* **1998**, *11*, 761–767. [CrossRef]

63. Jacobs, R.A. Increased rates of convergence through learning rate adaptation. *Neural Netw.* **1988**, *1*, 295–307. [CrossRef]

64. Jayalakshmi, T.; Santhakumaran, A. Statistical Normalization and Backpropagation for Classification. *Int. J. Comput. Theory Eng.* **2011**, *3*, 89–93. [CrossRef]

65. Neves Bez, H.; Yibole, H.; Pathak, A.; Mudryk, Y.; Pecharsky, V.K. Best practices in evaluation of the magnetocaloric effect from bulk magnetization measurements. *J. Magn. Magn. Mater.* **2018**, *458*, 301–309. [CrossRef]

66. Lozano, J.A.; Engelbrecht, K.; Bahl, C.R.H.; Nielsen, K.K.; Barbosa, J.R.; Prata, A.T.; Pryds, N. Experimental and numerical results of a high frequency rotating active magnetic refrigerator. *Int. J. Refrig.* **2014**, *37*, 92–98. [CrossRef]

Energies **2019**, *12*, 1871

67. Christiaanse, T.V.; Campbell, O.; Trevizoli, P.V.; Govindappa, P.; Niknia, I.; Teyber, R.; Rowe, A. A concise approach for building the s-T diagram for Mn-Fe-P-Si hysteretic magnetocaloric material. *J. Phys. D Appl. Phys.* **2017**, *50*, 365001. [CrossRef]
68. Mugica, I.; Poncet, S.; Bouchard, J. An open source DNS solver for the simulation of Active Magnetocaloric Regenerative cycles. *Appl. Therm. Eng.* **2018**, *141*, 600–616. [CrossRef]
69. Hansen, B.R.; Bahl, C.R.H.; Kuhn, L.T.; Smith, A.; Gschneidner, K.A.; Pecharsky, V.K. Consequences of the magnetocaloric effect on magnetometry measurements. *J. Appl. Phys.* **2010**, *108*, 043923. [CrossRef]

energies

MDPI

Article

CFD Simulation and Experimental Study of Working Process of Screw Refrigeration Compressor with R134a

Huagen Wu [1,*], Hao Huang [1,2,*], Beiyu Zhang [1], Baoshun Xiong [1] and Kanlong Lin [1]

[1]   School of Energy and Power Engineering, Xi'an Jiaotong University, Xi'an 710049, China;
     zby555544444@stu.xjtu.edu.cn (B.Z.); xiongbaoshun98@stu.xjtu.edu.cn (B.X.);
     longerme@stu.xjtu.edu.cn (K.L.)
[2]   State Key Laboratory of Compressor Technology, Hefei 230001, China
*   Correspondence: hgwu@xjtu.edu.cn (H.W.); xjhuanghao@stu.xjtu.edu.cn (H.H.);
     Tel.: +86-29-82664845 (H.W.)

Received: 2 April 2019; Accepted: 27 May 2019; Published: 29 May 2019

Abstract: Twin-screw refrigeration compressors have been widely used in many industry applications due to their unique advantages. The performance of twin-screw refrigeration compressors is generally predicted by one-dimensional numerical simulation or empirical methods; however, the above methods cannot obtain the distribution of the fluid pressure field and temperature field inside the compressor. In this paper, a three-dimensional model was established based on the experimental twin-screw refrigeration compressor. The internal flow field of the twin-screw compressor was simulated by computational fluid dynamics (CFD) software using structured dynamic grid technology. The flow and thermodynamic characteristics of the fluid inside the compressor were analyzed. The distribution of the internal pressure field, temperature field, and velocity field in the compressor were obtained. Comparing the P-θ indicator diagram and the performance parameters of the compressor with the experimental results, it was found that the results of the three-dimensional numerical simulation were consistent with the experimental data. The maximum error was up to 2.578% on the adiabatic efficiency at the partial load working condition. The accuracy of the 3D numerical simulation of the screw compressors was validated and a new method for predicting the performance of twin-screw refrigeration compressors was presented that will be helpful in their design.

Keywords: twin-screw refrigeration compressor; CFD; thermodynamic performance; P-θ indicator diagram

---

1. Introduction

Compared with other kinds of compressors, twin-screw refrigeration compressors have the characteristics of a simple structure, less easily damaged parts, good running stability, and strong adaptability. They also occupy a large proportion of the market. At present, they are been widely used in central air conditioning systems for commercial buildings and residential buildings. The study of the thermodynamic performance of twin-screw refrigeration compressors is helpful in optimizing the performance of the compressor and finding out the optimum operating condition of the compressor unit. The rotor profile is the most important part in the design of the screw compressor. The working process of the twin-screw compressor is completed by the periodic meshing rotation of the helical tooth surface of the male and female rotors. The profile of the rotor determines the leakage characteristics and dynamic characteristics of the screw compressor. Good sealing and reasonable torque distribution are the basic principles of the rotor profile design. Contact line length, closed volume, inter-tooth area, and leaking triangle are the basic elements to measure the pros and cons of the profile design, and are also closely related to the airflow pulsation and thermal performance of the compressor [1].

Wu et al. [2] studied the effect of lubricating oil on the performance of twin-screw refrigeration compressors. Through mathematical models and experimental studies, it was found that the lubricating oil injection position, the oil injection flow rate, and the oil filling temperature all affected the performance of the compressor. The research provides theoretical support for optimizing lubricating oil distribution and improving the efficiency of twin-screw refrigeration compressors. Through the experimental study of rotor axial force under different operating conditions, Hou [3] proposed the mean pressure model and sector pressure model to calculate the axial force on the rotor end face. It was concluded that the sector pressure model was more accurate in predicting the axial force on the rotors in comparison with that of the mean pressure model. Moreover, the axial force on the rotor end face was found to have a larger influence than that on the rotor helical surface. Many scholars have proposed mathematical models for the working process of twin-screw refrigeration compressors [4,5]. These models fully consider the various leakage paths of the compressor, oil and gas heat transfer, power loss, etc., and can effectively predict the performance of the compressor. However, these mathematical models do not provide detailed information on the fluid flow inside the compressor.

With the rapid development of computer technology, CFD is widely used in the research of compressor performance, which can not only reduce expensive experiment costs, but also effectively reduce development time and shorten the development cycle. Due to the complex rotor shape of the twin-screw compressor, there is a very small gap in the machine, which makes the grid generation very difficult. Kovacevic et al. [6,7] introduced an advanced grid generation method and developed a completely original boundary adaptive program that can be applied to any rotor shape. This method can improve the accuracy of the simulation calculation, can better predict the performance of the compressor, and allows the design of such machinery to have a lower development cost. Rane et al. [8,9] proposed a new algebraic technique for generating a deformed grid of a screw compressor, which has the advantages of algebraic methods and differential methods. It introduced two control functions to regularize the initial algebraic distribution, which greatly improved the quality and distribution of the grid unit and improved the motion stability of the grid nodes.

Willie et al. [10] simulated the flow field of the compressor when studying the noise source of the oil-free screw compressor on the truck, but the simulation results had higher errors than the experimental results. The possible reason is that it does not perform grid independence verification or the number of numerical cells is insufficient.

Rane et al. [11] studied a twin-screw compressor and found that the refinement of the grid along the circumferential direction of the rotor profile directly affected the prediction of mass flow. Ding and Kim et al. [12,13] used different CFD software to simulate the effect of oil injection on compressor performance, which could effectively reduce the discharge temperature and improve the performance of the compressor. Rane et al. [14] applied CFD technology to the study of screw compressors with variable pitch and variable rotor end faces. By changing the shape of the rotor, it could improve the compression characteristics and achieve the internal pressure of the compressor's rapid rise. At the same pressure ratio, the variable pitch and variable section rotors achieved a larger discharge area, and the variable pitch rotor could effectively reduce the length of the seal line in the high pressure region, thereby reducing leakage.

Byeon et al. [15] used overlapping grid technology to simulate the working process of an oil-free air compressor. The experiment obtained the adiabatic efficiency and volumetric efficiency of the compressor where the simulation results were very similar to the experimental results. Arjeneh [16] used the CFD method to study local pressure variation inside the suction plenum of the screw compressor, described the measurement method of related experiments, and verified the application of CFD technology in the research of screw machinery. Rane [17] studied the influence of CFD solvers on the performance prediction of a twin-screw compressor. The performance predictions obtained by calculations with these two CFD models were compared with measurements obtained on the test compressor. The study revealed differences between the results obtained by two different solvers and the experimental results and provides a reference for the choice of CFD solver.

This paper used CFD technology to simulate the working process of the twin-screw refrigeration compressor under different loads, analyze the flow characteristics of the compressor internal fluid, and obtain the pressure, velocity and temperature distribution. The simulation results were similar to the experimental data, which proved the model was correct. Studying the working characteristics of a twin-screw refrigeration compressor under partial load can provide a theoretical basis for its optimized design and avoid unnecessary power consumption.

2. Geometric Model Building and Meshing

The working process of the twin-screw compressor is very complicated. In order to facilitate the simulation and reflect the working process of the compressor, this study divided the fluid domain model into three parts: the suction port, the volumetric element, and the discharge port. The three parts needed to be connected in the calculation process, so that the data information could be exchanged. Figure 1 shows the control volume fluid domain of the compressor.

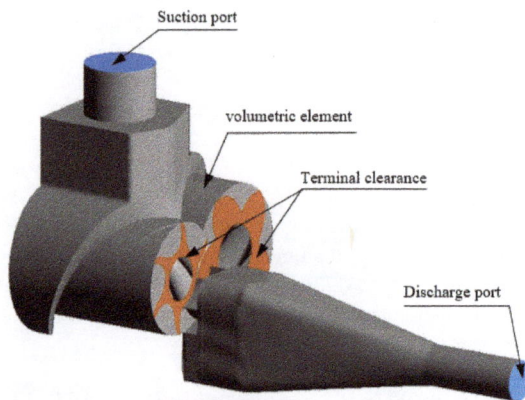

Figure 1. The control volume fluid domain of the compressor.

Since the shape of the suction and the discharge ports of the twin-screw refrigeration compressor is very irregular, structured hexahedral grids cannot be generated in ANSYS Mesh. Furthermore, the shape of the suction and the discharge ports does not change during the simulation. Therefore, the suction port and the discharge port were mesh blended by the tetrahedral grids and hexahedral grids as shown in Figure 2a,b. In order to ensure the accuracy of numerical interpolation during calculation, it was necessary to ensure that the mesh size of the interface between the suction port and the discharge port and the rotor cavity is the same. The mesh size of the interface was 1.25 mm, and the number of meshes of the suction port and the discharge port was 1,504,949. The mesh skewness was less than 0.81, and the quality met the calculation requirements.

Due to the complex geometry of the compressor's working chamber, the internal fluid domain changed continuously with the rotation of the rotor, and the existence of a small clearance, which made the requirements of the grid extremely demanding. Therefore, high-quality, fast mesh generation tools are needed to satisfy the solver's ability to accurately predict the boundary layer flow and flow within the clearance. This paper used TwinMesh to generate the fluid domain mesh in the working cavity. This is a mesh generation tool for the internal flow simulation of a positive displacement rotary compressor that can automatically generate a high quality hexahedral mesh according to the rotor profiles.

In TwinMesh, various parameters can be set to generate the grid. After grid independence verification, the number of circumferential nodes of the rotor profile was set to 720, and the maximum mesh size was set to 1 mm. It was necessary to draw a set of grids per degree, so a total of 72 sets of grids needed to be generated. When simulating, ANSYS-CFX can automatically select the corresponding grid for calculation. In order to make up for the oil's sealing, which will reduce the leakage of refrigerant, the leakage clearance of the compressor was appropriately reduced when generating the cells in TwinMesh, as shown in Figure 2c. The minimum angle of the grid needed to be greater than 18° to meet the calculation requirements. The closer the grid angle is to 90°, the better the grid quality. Finally, the number of hexahedral meshes generated in the volumetric element fluid domain was 5,279,175, and the quality met the requirements.

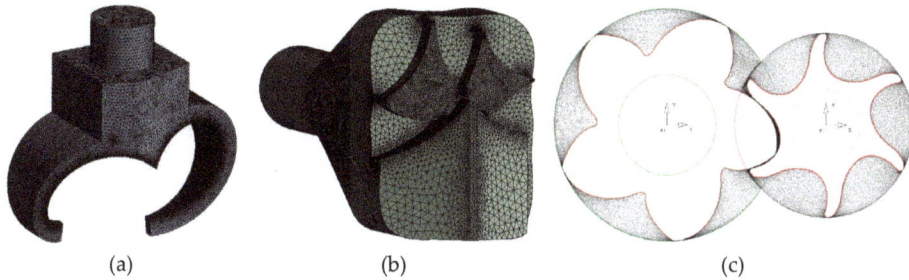

(a) (b) (c)

Figure 2. Meshing of different fluid domains. (**a**) Suction port fluid domain grid. (**b**) Discharge port fluid domain grid, (**c**) Two-dimensional mesh of volumetric element.

3. Simulation Settings

3.1. Governing Equation

The flow and convective heat transfer of fluids are controlled by the laws of conservation of physics, which include the law of conservation of mass, the law of conservation of momentum, and the law of energy conservation. Since the flow is in a turbulent motion state, the system also needs to comply with the additional turbulent transport equation. The governing equations are mathematical descriptions of these conservation laws.

Any flow problem must satisfy the law of conservation of mass, and the mass added to the fluid microelement in unit time is equal to the net mass flowing into the microelement at the same time. According to this law, a general form of mass conservation equation can be obtained, which is applicable to compressible flow and incompressible flow. The source term S_m can be any custom source term:

$$\frac{\partial \rho}{\partial t} + \frac{\partial}{\partial x_i}(\rho u_i) = S_m. \tag{1}$$

The law of conservation of momentum is also a law that must be satisfied by any flow system. The rate of change of the momentum of a fluid in a microelement to time is equal to the sum of the forces acting on the microelement:

$$\begin{cases} \frac{\partial(\rho u)}{\partial t} + div(\rho u \vec{u}) = -\frac{\partial p}{\partial x} + \frac{\partial \tau_{xx}}{\partial x} + \frac{\partial \tau_{yx}}{\partial y} + \frac{\partial \tau_{zx}}{\partial z} + F_x \\ \frac{\partial(\rho v)}{\partial t} + div(\rho v \vec{u}) = -\frac{\partial p}{\partial x} + \frac{\partial \tau_{xy}}{\partial x} + \frac{\partial \tau_{yy}}{\partial y} + \frac{\partial \tau_{zy}}{\partial z} + F_y \\ \frac{\partial(\rho w)}{\partial t} + div(\rho w \vec{u}) = -\frac{\partial p}{\partial z} + \frac{\partial \tau_{xz}}{\partial x} + \frac{\partial \tau_{yz}}{\partial y} + \frac{\partial \tau_{zz}}{\partial z} + F_z \end{cases} \tag{2}$$

A flow system containing heat exchange must also satisfy the law of conservation of energy; it is the necessary equation for obtaining the temperature field of the fluid. The increase rate of energy in the microelement is equal to the net heat flow into the microelement plus the work done by the volume force and surface force on the microelement. The essence of this law is the first law of thermodynamics:

$$\frac{\partial(\rho T)}{\partial t} + div(\rho u T) = div\left(\frac{k}{c_p} gradT\right) + S_T. \tag{3}$$

Since the flow is in a turbulent motion state, the system also needs to comply with the additional turbulent transport equation to solve. This paper used the Shear Stress Transport model. The shear stress transport (SST) model is improved by the standard k-ω model, which has higher precision and reliability:

$$\begin{cases} \frac{\partial(\rho k)}{\partial t} + \frac{\partial(\rho k u_i)}{\partial x_i} = \frac{\partial}{\partial x_j}\left(\Gamma_k \frac{\partial k}{\partial x_j}\right) + G_k - Y_k + S_k \\ \frac{\partial(\rho \omega)}{\partial t} + \frac{\partial(\rho \omega u_i)}{\partial x_i} = \frac{\partial}{\partial x_j}\left(\Gamma_\omega \frac{\partial \omega}{\partial x_j}\right) + G_\omega - Y_\omega + D_\omega + S_\omega \end{cases} . \tag{4}$$

3.2. Boundary Condition Settings

The refrigerant in the compressor was a compressible fluid. The turbulence model in this paper was the SST k-ω model, which combines the advantages of the k-ω model in the near-wall region calculation and the advantages of the k-ε model in the far-field calculation. Considering the heat transfer caused by fluid flow, this paper used the total energy model to calculate the convection heat transfer and heat conduction, which was suitable for the heat transfer calculation of high-speed flow and compressible flow. The refrigerant was R134a, and its physical properties were given by the physical property data fitted by the Peng–Robinson equation in the database. As shown in Table 1, the inlet and outlet temperatures and pressures obtained from the experiment were used as the boundary conditions for CFD simulation. Since the discharge temperature of R134a is low, before entering into the compressor, the oil was not cooled in the experimental research, thus the cooling effect of the oil would be cut down greatly. Based on that the heat transfer of the oil was not considered in the CFD simulation. The rotor speed was 2960 rpm.

Table 1. Boundary conditions of computational fluid dynamics (CFD) simulation.

Parameters	Measured Value
Inlet pressure	0.285 MPa
Inlet temperature	274 K
Discharge pressure	1.515 MPa
Discharge temperature	360.4 K
Volume flow	3.254 $m^3 \cdot min^{-1}$

3.3. Grid Independence Verification

Since the number of cells will affect the simulation results, this paper used meshes of different parameter settings in TwinMesh for the simulation. Figure 3 shows the control volume cells of the experimental compressor. The simulation results under different parameters are shown in Table 2.

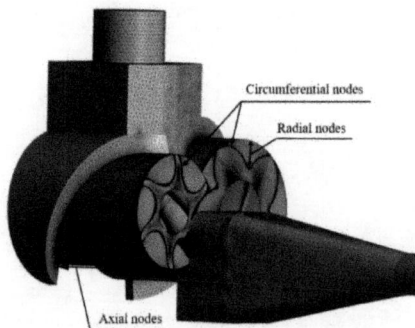

Circumferential nodes

Radial nodes

Axial nodes

Figure 3. The control volume cells of the experimental compressor.

Table 2. Grid independence verification.

Case	Circumferential Nodes	Radial Nodes	Axial Nodes	Volume Flow/m^3·min^{-1}	Number of Meshes
Case 1	360	30	125	2.633	3,848,194
Case 2	720	20	125	2.807	4,536,394
Case 3	720	30	125	2.837	6,477,694
Case 4	720	20	200	3.248	6,693,394
Case 5	720	20	250	3.251	8,203,294

Figure 4a shows the cell numbers in different cases, Figure 4b shows the volume flow in different cases, Figure 4c shows the P-θ diagram in different cases, and Figure 4d shows the power in different cases. It can be seen from the figure that the number of circumferential nodes of the rotor profile and the number of axial nodes had a great influence on the compressor volume flow, and the number of radial nodes had little influence on the compressor volume flow. The number of cells had the greatest influence on the volume flow, and had little effect on the pressure and the indicated power, but the volume flow is one of the most important performance parameters of the compressor. Therefore, the volume flow should be used as the basis for the simulation results. Considering the cost of computing resources, this paper used the mesh of Case4 for the simulation calculations.

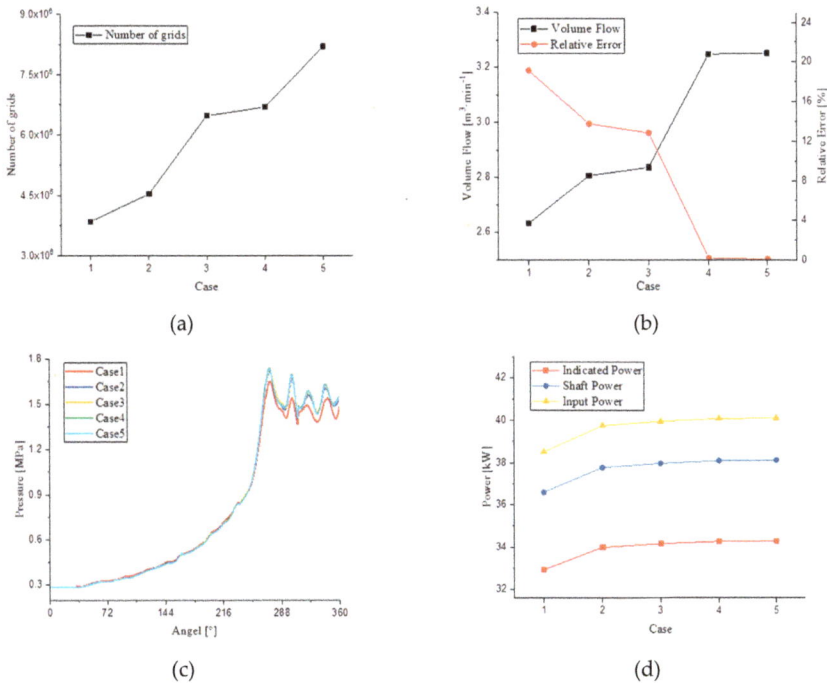

(a)

(b)

(c)

(d)

Figure 4. Mesh independence verification. (**a**) Mesh numbers in different cases. (**b**) Volume flow in different cases. (**c**) P-θ diagram in different cases. (**d**) Power in different cases.

4. Experiment

In general, the volumetric flow of a twin-screw refrigeration compressor needs to be designed according to the maximum gas consumption. Due to various external climatic conditions, different applications, and other reasons, twin-screw refrigeration compressors have been under partial load conditions for a long time. Studying the working characteristics of a twin-screw refrigeration compressor under partial load can provide a theoretical basis for its optimized design and avoid unnecessary power consumption.

Based on the working characteristics of twin-screw refrigeration compressors, the slide valve adjusting device driven by hydraulic is commonly used to realize the capacity adjustment process. The body of the twin-screw refrigeration compressor is equipped with a slide valve with a radial discharge port. When there is a bypass port between the low-pressure end of the slide valve and the fixed block, the effective working length of the rotors is reduced, and the compressor is in a partial load state. In this paper, the 75% load condition refers to where the volume flow is 75% of the maximum volume flow, and the 100% load condition refers to when the volume flow is the maximum volume flow.

In order to verify the correctness of the CFD simulation, a micro pressure sensor was installed near the discharge end face in a tooth groove of the female rotor that could monitor the pressure signal in a working cycle, so that the P-θ diagram of the compressor in the working process could be obtained. The pressure sensor (XT-140-250A) used in this paper had the characteristics of non-linearity, repeatability, hysteresis not exceeding 5%, and high resolution. Its maximum working pressure was 3.5 MPa, its natural frequency was 700 Hz, and it could be effectively used in the temperature range of −50–205 °C. The modified female rotor is shown in Figure 5a where the micro pressure sensor installed in the female rotor tooth groove collected the pressure signal in the tooth groove. The pressure signal was transmitted through the signal line located in the female rotor axis. The signal was transmitted to the moving plate of the collecting ring. Next, the collecting ring converted the pressure signal to the static plate and sent it to the signal amplifier. The pressure signal was amplified by the signal amplifier and then entered the dynamic signal analyzer for data acquisition. A corner marking disc with a slit was fixed on the female rotor. When the disc rotated at a high speed, the photoelectric sensor periodically sensed the light transmitted from the slit and generated a pulse signal, and the pulse signal was used as the mark of the position of the female rotor. The pressure signal and the pulse signal entered the dynamic signal analyzer at the same time, and the P-θ diagram was obtained after being processed.

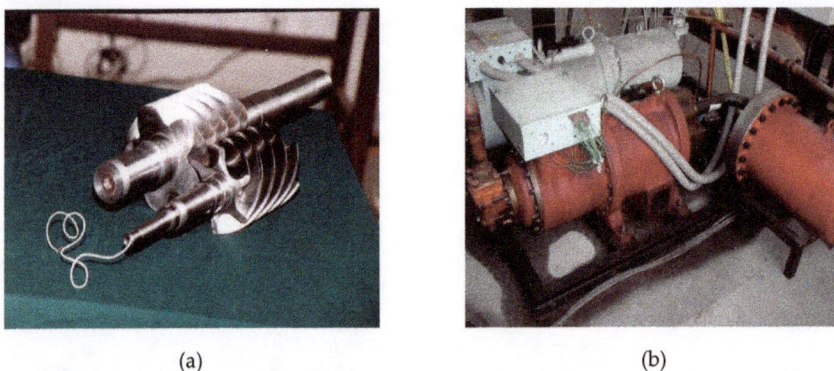

| (a) | (b) |

Figure 5. Experimental device. (**a**) Pressure sensor in the female rotor. (**b**) Experimental test compressor.

Figure 5b shows the experimental test compressor. In order to obtain the P-θ diagram of the screw compressor, it was necessary to modify the existing screw compressor to meet the requirements of the test. The experiment was carried out under partial load (75%) and full load.

Table 3 shows the detailed structural parameters of the twin screw refrigeration compressor used in the experiment. The ratio of the rotor length to the outside diameter of the male rotor (L/D) was 1.09742. The theoretical volume flow rate of the compressor was 3.797 $m^3 \cdot min^{-1}$.

Table 3. Structural parameters of the screw compressor.

Parameter	Male Rotor	Female Rotor
Number of teeth	5	6
Diameter of rotor D/mm	138.507	109.76
Length of rotor L/mm	152	152
Wrap angle of male rotor	300°	250°
Screw lead of the male rotor T/mm	181.761	218.113

5. Results and Discussion

The effect of the pressure distribution inside the compressor on the efficiency and performance of the compressor is a topic for engineers in the compressor field. The distribution of pressure in the compressor was calculated by the CFD software and the variation of gas pressure inside the flow channel was analyzed in detail.

As the pressure on the rotor surface represents the distribution of pressure in the working volume of the compressor, the pressure changes inside the compressor at different angles are shown in Figure 6. Since a 5-6 rotor profile was used in this simulation, the pressure changed periodically every 72°. Therefore, the pressure distribution at the rotation angles of 0° and 72° was identical, and the pressure distribution at 18° and 90° was also the same. The pressure gradually increased along the spiral surface of the rotor from the suction port to the discharge port, and the pressure of the discharge port reached the maximum.

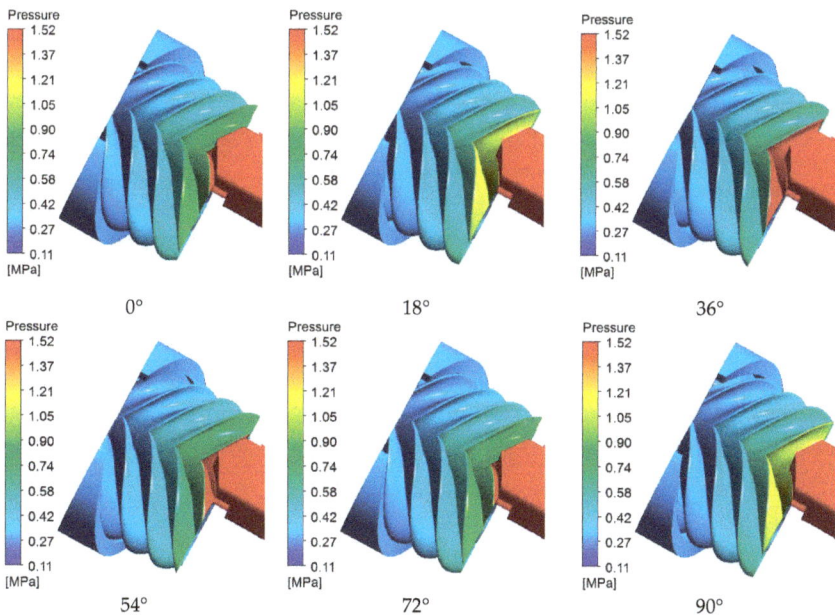

Figure 6. The pressure distribution inside the compressor at different angles.

Figure 7 shows the distribution of pressure in the mesh area with the male rotor and female rotor. It can be seen from Figure 7 that the pressure on the rotor surface was obviously divided into two parts: the high pressure area and the low pressure area of the suction pressure, which was caused by the contact line between the rotors that meshed with each other. Figure 8 shows the pressure variation when the gas flows through the intermeshing clearance at different angles. There is an area in the low-pressure region close to the contact line that has a lower pressure than the suction pressure, and becomes more and more obvious along the suction end face to the discharge end face. This could be due to the fact that the intermeshing clearance between the rotors along the contact line is too small, resulting in an increasingly pronounced effect of throttling as the differential pressure increases.

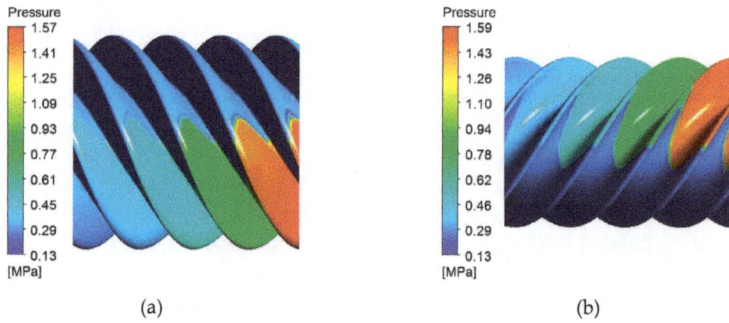

(a)   (b)

Figure 7. The pressure distribution of the rotors surface. (**a**) Male rotor; (**b**) Female rotor.

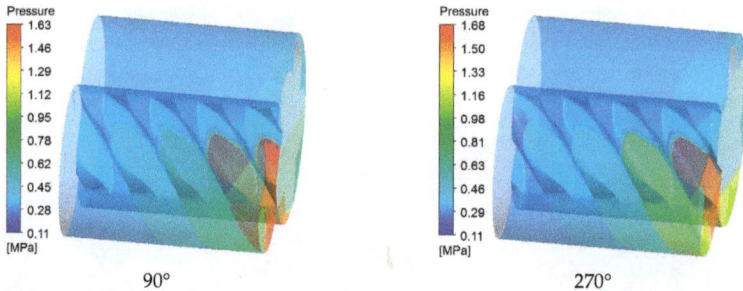

90°   270°

Figure 8. The pressure distribution of the intermeshing clearance of the rotors at different angles.

Inside the compressor, the temperature rise is mainly caused by the following parts: the heat generated by the friction between gas and casing, the heat generated by the compressed gas, and the flow loss and impact loss. Figure 9 shows the temperature distribution of the surface of the male and female rotors. Figure 10 shows the temperature distribution inside the compressor at different angles. It can be seen that the temperature gradually rose along the axial direction of the rotor, and the temperature near the discharge port was the highest. Additionally, the temperature at the discharge port was significantly larger than the suction port, which was in line with the actual situation. This part of the study can provide a reference for the location design of the oil injection port.

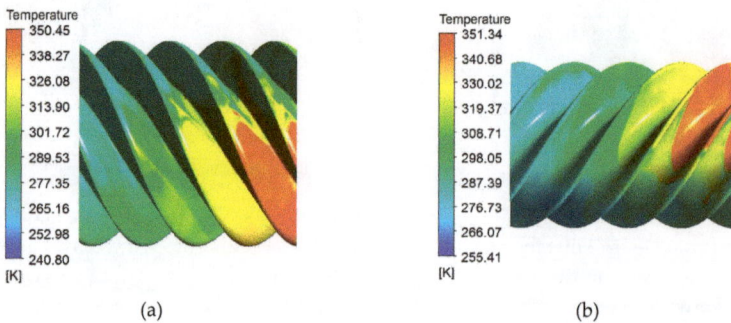

(a)   (b)

Figure 9. The temperature distribution of the surface of the male and female rotors: (**a**) male rotor; (**b**) female rotor.

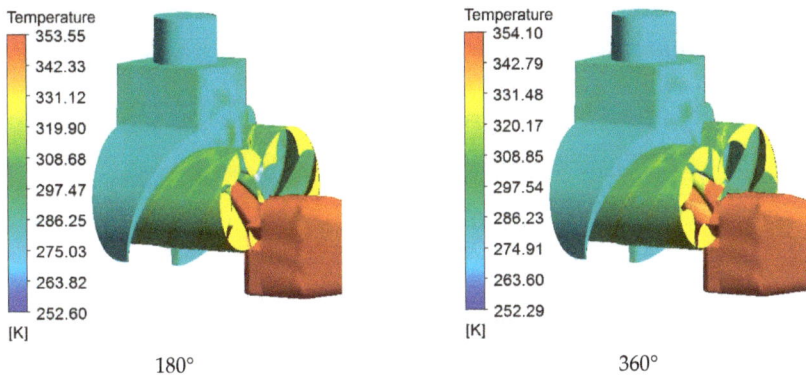

Figure 10. The temperature distribution inside the compressor at different angles.

As shown in Figure 11, due to the intermeshing clearance between the male and female rotors, the gas flows back through the intermeshing clearance during the high-speed rotation of the rotor, causing the leakage from the high-pressure chamber to the low-pressure chamber. The velocity in the leakage increases as the pressure differential across the adjacent volumetric elements of the rotor becomes higher, and the velocity of the fluid at the intermeshing clearance is much greater than the velocity of the fluid in the volumetric element. For the same reason, the radial clearance of the compressor also has this leakage problem. Figure 12 shows the leakage velocity vector of the compressor at different angles. It can be seen that the refrigerant flow rate between the different volumetric elements hardly changed and the leakage in the terminal clearance was obvious. The maximum leakage velocity exceeded 200 m/s. The leakage of gas in the compressor was mainly caused by the difference in pressure between the different slots, and the distribution law of the leakage velocity was positively correlated with the pressure difference. Gas leakage changes the flow state of the internal airflow of the compressor and greatly reduces the efficiency of the compressor, so it is especially necessary to properly control the leakage of the twin-screw compressor.

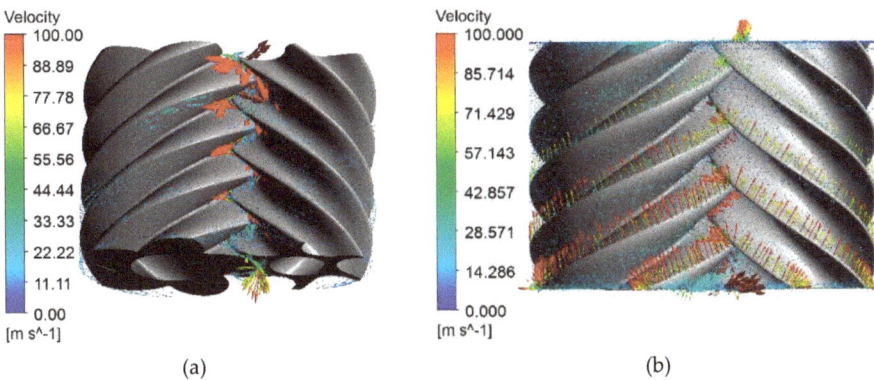

(a)   (b)

Figure 11. The leakage velocity vector: (**a**) the leakage velocity vector of the intermeshing clearance; (**b**) the leakage velocity vector of the radial clearance.

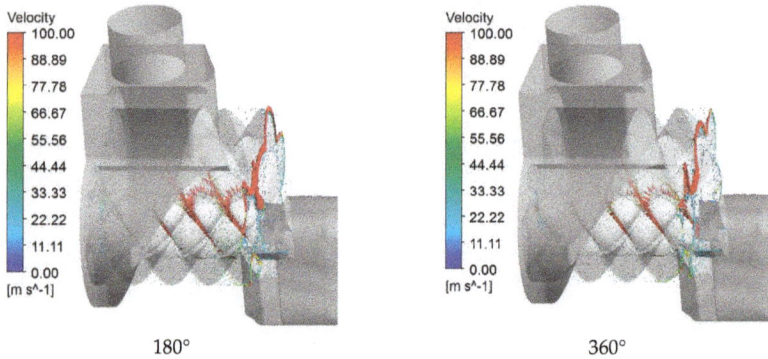

180°   360°

Figure 12. The leakage velocity vector of the compressor at different angles.

This paper verified the correctness of the CFD model from two aspects: First, the P-θ diagram of the working process of the screw refrigeration compressor obtained by CFD simulation was compared with that obtained by the experiments, and the pressure in the working chamber of the compressor constantly changed with the angle of the male rotor, which was to construct important parameters of the P-θ diagram. If the P-θ diagram calculated by the CFD agrees well with the experimentally measured P-θ diagram, it indicates that the prediction of compressor performance by this CFD calculation model is correct. Second, performance parameters such as volumetric efficiency, isentropic efficiency, and input power of the compressor obtained by CFD simulation were compared with that obtained by the experiments.

Figures 13 and 14 show the P-θ diagram of the screw refrigeration compressor under partial load (75%) and full load conditions. It can be seen that the pressure curve obtained by the CFD calculation agreed well with the experimentally measured pressure curve. In the discharge process, the CFD calculation model failed to accurately reflect the actual process, which may have been caused by the large pressure pulsation in the discharge process.

As can be seen in Figure 13, the starting point of the compression process was delayed, the compression curve became steep, and the discharge time became shorter. At the beginning of the compression process, the capacity control slide valve causes the compress part to be connected to the discharge part, which causes the compression process to be delayed. This causes the compressor discharge pressure to be slightly lower than the compressor discharge pressure at full load. Due to the change in the radial discharge orifice, the compressor's discharge time was also slightly less than the compressor's discharge time under a full load.

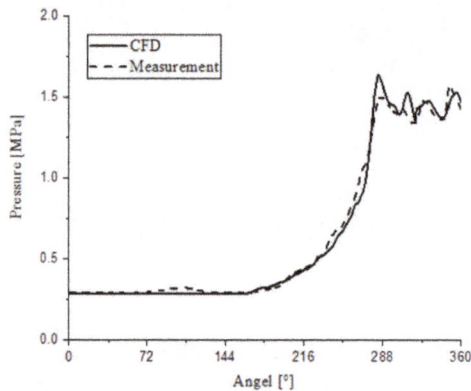

Figure 13. The P-θ diagram of the screw refrigeration compressor under partial load (75%) conditions.

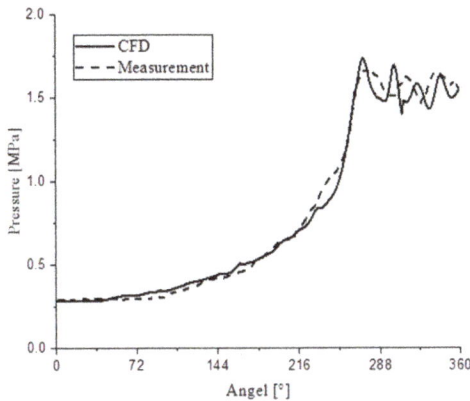

Figure 14. The P-θ diagram of the screw refrigeration compressor under full load conditions.

In the CFD calculation, the indicated power of the compressor under partial load conditions was 22.758 kW, and the indicated power of the compressor under full load conditions was 34.271 kW. The motor efficiency was 0.95 and the mechanical efficiency was 0.91. The input power was 26.325 kW and 39.643 kW, respectively. Table 4 shows the comparison of the macroscopic performance of the compressor under two working conditions. It can be seen from the table that the volumetric efficiency, isentropic efficiency, and input power of the compressor calculated by CFD were in good agreement with the experimentally measured data.

Table 4. Comparison of the macroscopic performance of the compressor under two working conditions.

Method	75% Load			100% Load		
	Input Power	Volumetric Efficiency/%	Isentropic Efficiency/%	Input Power	Volumetric Efficiency/%	Isentropic Efficiency/%
Measured	26.986	52.583	61.281	39.9	85.699	70.356
CFD	26.325	51.848	59.701	39.643	85.541	69.686
Error%	2.449	1.398	2.578	0.645	0.184	0.952

As shown in Figure 15, under full load conditions, the error between the results obtained by the CFD calculation and the experimental results was less than 1%. Under partial load conditions, the error of input power and isentropic efficiency calculated by CFD was large, but the error value was less than 3%. The results of both conditions were within the tolerances.

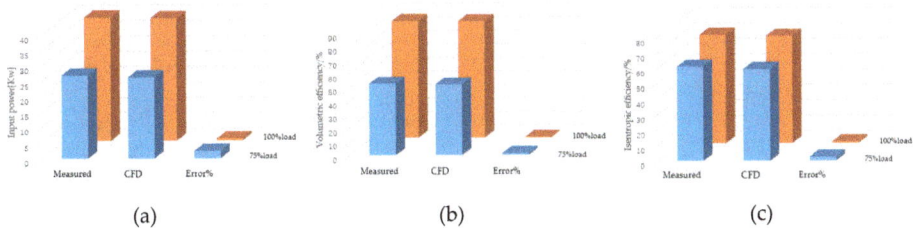

Figure 15. Comparison of the compressor performance under two working conditions. (**a**) Input power; (**b**) Volumetric efficiency; (**c**) Isentropic efficiency.

The accuracy of the CFD calculation model was fully illustrated by the comparison between the CFD calculation results and the experimental results. This model can correctly describe the working processes of the compressor.

6. Conclusions

In this paper, the internal flow field of the twin-screw compressor was numerically simulated by the fluid analysis software CFX, and the flow characteristics of the internal flow field of the compressor were obtained. Our results can be concluded as follows.

The distribution of the pressure, the temperature, and the speed in the leakage clearance inside the compressor were visualized. The gas temperature from the suction port to the discharge port of the compressor was greatly improved, and the temperature was highest at the discharge port. It could be clearly observed that the pressure at the contact line of rotors was the lowest. This could be due to the fact that the intermeshing clearance between the rotors along the contact line was too small, resulting in an increasingly remarkable influence of throttling as the pressure difference increased between the high-pressure chamber and the low-pressure chamber. The refrigerant leakage velocity at the compressor clearance was larger than that in the tooth grooves. The leakage was most obvious from the high-pressure chamber to the low-pressure chamber. The maximum leakage velocity of the clearance exceeded 200 m/s. Therefore, various leakage clearance should be minimized during the design of the compressor.

On the basis of the experimental investigation, the P-θ diagram of the twin-screw refrigeration compressor under different load conditions was obtained, and the performance such as input power, volumetric efficiency, and adiabatic efficiency of the compressor were tested. By comparing the CFD calculation results with the experimental results, it was found that the CFD model of this paper could accurately show the working process of the twin-screw compressor, and the parameters such as discharge temperature and pressure were close to the test results. This provides a reliable method for the design and optimization of twin-screw refrigeration compressors.

Author Contributions: Investigation, H.W. and H.H.; Writing—review & editing, B.Z., B.X. and K.L.

Funding: This research was funded by the National Natural Foundation of China, grant number 50806055 and State Key Laboratory of Compressor Technology Open Fund Project, grant number SKL-YS.1201805.

Conflicts of Interest: The authors declare no conflict of interest.

Nomenclature

S_m	Source term
p	Pressure on the microelement
τ_{xy}	Component of the viscous stress τ
F	Volume force on the microelement
K	Heat transfer coefficient
T	Temperature
c_p	Specific heat capacity
S_T	Internal heat source and the viscous dissipation term
G_k	Kinetic energy of turbulence
Γ_k	Effective diffusion terms of k
Γ_ω	Effective diffusion terms of ω
Y_k	Divergent terms for k
Y_ω	Divergent terms for ω
D_ω	Orthogonal divergence
$S_k \& S_\omega$	User-defined item

References

1. Hauser, J.; Brümmer, A. Geometrical abstraction of screw compressors for thermodynamic optimization. *Proc. Inst. Mech. Eng. Part C J. Mech. Eng. Sci.* **2011**, *225*, 1399–1406. [CrossRef]
2. Wu, X.; Xing, Z.; He, Z.; Wang, X.; Chen, W. Effects of lubricating oil on the performance of a semi-hermetic twin screw refrigeration compressor. *Appl. Therm. Eng.* **2017**, *112*, 340–351. [CrossRef]

3. Hou, F.; Zhao, Z.; Yu, Z.; Xing, Z. Experimental study of the axial force on the rotors in a twin-screw refrigeration compressor. *Int. J. Refrig.* **2017**, *75*, 155–163. [CrossRef]
4. Giuffrida, A. A semi-empirical method for assessing the performance of an open-drive screw refrigeration compressor. *Appl. Therm. Eng.* **2016**, *93*, 813–823. [CrossRef]
5. Wu, H.; Xing, Z.; Shu, P. Theoretical and experimental study on indicator diagram of twin screw refrigeration compressor. *Int. J. Refrig.* **2004**, *27*, 331–338.
6. Kovacevic, A. Boundary adaptation in grid generation for CFD analysis of screw compressors. *Int. J. Numer. Methods Eng.* **2005**, *64*, 401–426. [CrossRef]
7. Kovacevic, A.; Stosic, N.; Smith, I. Grid generation of screw machine geometry. In *Screw Compressors*; Springer: Berlin/Heidelberg, Germany, 2007.
8. Rane, S.; Kovacevic, A. Algebraic generation of single domain computational grid for twin screw machines. Part I. Implementation. *Adv. Eng. Softw.* **2017**, *107*, 38–50. [CrossRef]
9. Rane, S.; Kovacevic, A. Application of numerical grid generation for improved CFD analysis of multiphase screw machines. *IOP Conf. Ser. Mater. Sci. Eng.* **2017**, *232*. [CrossRef]
10. Willie, J. Use of CFD to predict trapped gas excitation as source of vibration and noise in screw compressors. *IOP Conf. Ser. Mater. Sci. Eng.* **2017**, *232*. [CrossRef]
11. Rane, S.; Kovacevic, A.; Stosic, N.; Kethidi, M. Deforming grid generation and CFD analysis of variable geometry screw compressors. *Computers Fluids* **2014**, *99*, 124–141. [CrossRef]
12. Ding, H.; Jiang, Y. CFD simulation of a screw compressor with oil injection. *IOP Conf. Ser. Mater. Sci. Eng.* **2017**, *232*. [CrossRef]
13. Kim, Y.J.; Byeon, S.; Lee, J.Y. Numerical analysis on the flow characteristics of oil-injected screw air compressor. In Proceedings of the 2014 ISFMFE—6th International Symposium on Fluid Machinery and Fluid Engineering, Wuhan, China, 22–25 October 2014.
14. Rane, S.; Kovacevic, A.; Stosic, N.; Kethidi, M. CFD grid generation and analysis of screw compressor with variable geometry rotors. In Proceedings of the 8th International Conference on Compressors and their Systems, London, UK, 9–10 September 2013.
15. Byeon, S.-S.; Lee, J.-Y.; Kim, Y.-J. Performance Characteristics of a 4 x 6 Oil-Free Twin-Screw Compressor. *Energies* **2017**, *10*, 945. [CrossRef]
16. Arjeneh, M.; Kovacevic, A.; Rane, S.; Manolis, M.; Stosic, N. Numerical and experimental investigation of pressure losses at suction of a twin screw compressor. *IOP Conf. Ser. Mater. Sci. Eng.* **2015**, *90*. [CrossRef]
17. Kovacevic, A.; Rane, S.; Stosic, N.; Jiang, Y.; Lowry, S.; Furmanczyk, M. Influence of Approaches in CFD Solvers on Performance Prediction in Screw Compressors. Available online: https://docs.lib.purdue.edu/icec/2252/ (accessed on 15 January 2019).

energies

MDPI

Article

Modeling and Optimizing a Chiller System Using a Machine Learning Algorithm

Jee-Heon Kim [1], Nam-Chul Seong [1] and Wonchang Choi [2],*

[1] Eco-System Research Center, Gachon University, Seongnam 13120, Korea
[2] Department of Architectural Engineering, Gachon University, Seongnam 13120, Korea
* Correspondence: wchoi@gachon.ac.kr; Tel.: +82-31-750-5335

Received: 12 June 2019; Accepted: 24 July 2019; Published: 25 July 2019

Abstract: This study was conducted to develop an energy consumption model of a chiller in a heating, ventilation, and air conditioning system using a machine learning algorithm based on artificial neural networks. The proposed chiller energy consumption model was evaluated for accuracy in terms of input layers that include the number of input variables, amount (proportion) of training data, and number of neurons. A standardized reference building was also modeled to generate operational data for the chiller system during extended cooling periods (warm weather months). The prediction accuracy of the chiller's energy consumption was improved by increasing the number of input variables and adjusting the proportion of training data. By contrast, the effect of the number of neurons on the prediction accuracy was insignificant. The developed chiller model was able to predict energy consumption with 99.07% accuracy based on eight input variables, 60% training data, and 12 neurons.

Keywords: chiller energy consumption; artificial neural network (ANN); HVAC

---

1. Introduction

The energy consumption of a building system can be controlled by employing an energy-saving design as well as by proper operation of the building. Therefore, such design, referred to as a building energy management system (BEMS), has been adopted for building operations to ensure effective energy consumption and management. The integrated measurement, control, management, and operation in a BEMS provide efficient energy management and allow the desired indoor environment to be maintained. However, currently a BEMS is limited to a simple on/off switch that allows comparisons of the actual value with a set value determined by the operator. Rational energy management tools should be able to perform sophisticated functions that help managers make good BEMS decisions. Collection and analysis of available data are necessary in order to develop and provide such advanced management tools.

Recently, machine learning algorithms have been actively applied to optimize new type of heating, ventilation, and air conditioning (HVAC) systems [1–6]. For example, Jang et al. [7] proposed a solution for predicting the optimal heating time in winter by using artificial neuron network (ANN) technology. For this type of application, ANN models utilize input data such as the indoor/outdoor temperature, indoor/outdoor temperature difference, indoor temperature change, and outdoor temperature change. Jeong et al. [8] compared the accuracy of different building energy consumption predictions obtained by using three machine learning algorithms: ANN, a support vector machine, and random forest (RF). The performance was ranked in the order of RF, ANN, and support vector machine. For their study, Jeong et al. [9] investigated the sensitivity and importance of these variables for predicting the energy consumption in elementary schools and commercial buildings and then evaluated the performance of the machine learning models according to building function. For the elementary school buildings, the average coefficient of variance of the root mean squared error (CvRMSE) determined by ANNs was 5.4% and the average CvRMSE for commercial buildings based on the RF model was 10.9%.

Jeon et al. [10] conducted a study of energy load predictions for a building unit. Their study used an ANN and a test reference year. For each training period, the average CvRMSE for the prediction of the energy load was 25%. Park et al. [11] proposed an ANN model that can predict the cooling load according to the setback temperature in order to minimize the cooling energy consumption. Their results confirm a CvRMSE of 21.3%, which reflects a better performance than the conventional criterion of 30.0%. Ahmed et al. [12] investigated the performance of power load predictions associated with ANN and RF algorithms for a single building unit using meteorological data. Their ANN model obtained an average CvRMSE of 4.9% through normalization, extraction, and elimination of input variables to improve prediction performance. The RF model led to an average CvRMSE of 6.1%.

Seong et al. [13] developed and verified a building energy prediction model based on time series auto-regressions artificial neural networks based on the input variables of dry bulb and outdoor air temperature, hour of day, and type of day. As a result, CvRMSE was 40.9% for the numerical analysis model and 28.3% for the ANN model, which improved the accuracy. Seong et al. [14] developed an artificial neural network-based air flow prediction model to observe changes in accuracy with the number of input values. As a result, it was found that the predicted performance improved significantly as the number of inputs increased. Kim et al. [15] developed an artificial neural network-based energy consumption prediction model for fans to evaluate the accuracy according to input conditions. Mean bias error (MBE) showed a distribution of 1.7% to 2.95% of the learning period and 2.3% to 4.5% of the utilization period, while CvRMSE showed high predictive accuracy as it was distributed of 2.9% to 4.4% of the learning period and 3.6% to 7.9% of the utilization period.

There are many attempts to reduce energy consumption for the chiller systems including by using multiple linear regression analysis, interaction analysis of each components, and data-driven analysis [16–18].

In sum, various studies of machine learning methods, including ANN models, have been conducted in the field of building energy. The estimation of the energy consumption of a building, cooling load, etc. has been studied with results of high accuracy. However, most of the previous research has focused on the entire building system, and thus, management tools such as a BEMS need to be able to predict also the energy usage of subsystems such as air conditioners, heat source equipment, and transportation equip. To that end, the authors have developed management software for a centralized HVAC system. As shown in Figure 1, the centralized HVAC management software is composed of a real-time operation and performance monitoring function, energy performance prediction and optimization application, a performance evaluation report, and all the functions that utilize real-time data collected from a BEMS. Figure 1 shows the schematic diagram of a centralized HVAC management software, energy performance prediction and optimization function (EPPOF).

The developed software, referred to as energy performance prediction and optimization function (EPPOF), uses machine learning techniques to calculate control variables for predicting and optimizing energy consumption in a centralized HVAC system. To predict the energy consumption of the system, a prediction model based on ANNs was developed for the air conditioner, air handling unit, and heat source equipment that constitute the main energy consumers in an air conditioning system.

In this study, the authors investigated the accuracy of this developed model that is based on ANNs with respect to the input parameters, i.e., the number of input variables, proportion of training data, and number of neurons. The generated chiller operational data were generated and used to evaluate correlation between possible parameters in the chiller model. Section 2 presents the predictive energy consumption model for HVAC system with ANN. In Section 3, the accuracy for the predicted models with respect to input variations was evaluated and then Section 4 predicts the prediction of energy consumption based on the optimized input variations. Section 5 shows the conclusion based on the comparison between the simulated results and predicted results.

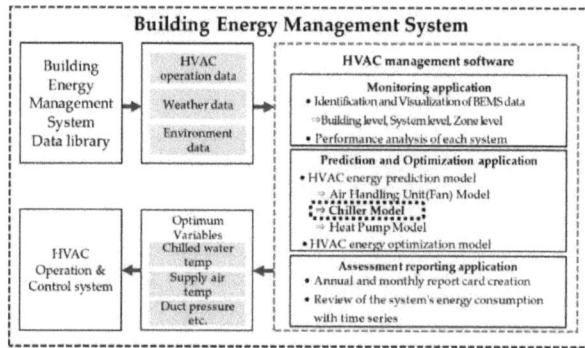

Figure 1. Schematic diagram of centralized heating, ventilation, and air conditioning (HVAC) management software, energy performance prediction and optimization function (EPPOF).

2. Predictive Model for Energy Consumption of a Chiller in an HVAC System

2.1. Modeling of the Reference Building Unit

A large amount of data is required for both training and testing with ANNs. For this study, the authors modeled a large-scale office building, designated in the commercial prototype building models in the U.S. Department of Energy (DOE) Building Energy Codes Program [19], and modified the related input variables for comparable buildings in Korea. Specifically, the climate data for Seoul, Korea were collected using the TRY (test, reference, year) format. This study generated the necessary data in accordance with the requirements for the DOE reference building model that has a standard pattern for energy consumption (annual energy consumption per area). Several parameters were considered to generate the reference building model: The heat source equipment, HVAC equipment, and energy performance variables, including core size, roof type, structure, construction year, heat flow rate, window area ratio, etc. Energy simulation software (Energyplus) was used to generate the related chiller operational data. The generated building was 12 storeys and a basement with floor area of 46320 m², and rectangular shape with aspect ratio of 1.5. Table 1 presents detailed boundary condition associated with reference building.

Table 1. Simulation condition of reference building large scale office.

Component	Features
Weather Data and Site Location	TRY Seoul (latitude: 37.57°N, longitude: 126.97°E)
Building Type	Large Scale Office
Total Building Area (m²)	46320
Hours Simulated (hour)	3761
Envelope Insulation (m²K/W)	External Wall 0.35, Roof 0.213, External Window 1.5
Window-Wall Ratio (%)	40
Set Point (°C)	Cooling 26, Heating 20
Internal Gain	Lighting 10.76 (W/m²), People 18.58 (m²/person), Plug and Process 10.76 (W/m²)
HVAC Sizing	Auto Calculated
HVAC Operation Schedule	7:00–18:00

Table 2 presents the specification chiller system used in this study.

Table 2. Chiller specification.

Type	Nominal Capacity	Nominal Efficiency	Initial Design Size Reference Chilled Water Flow Rate	Design Size Reference Chilled Water Flow Rate	Design Size Reference Condenser Water Flow Rate	Design Chilled Water Temperature
Electric:EIR	5114517 W	5.5 W/W	0.12 m³/s	0.18 m³/s	0.26 m³/s	6.7 °C

2.2. Determination of Input Values

Among the many variables in the dataset that were generated for the reference building model, some were selected to be used as input values for the ANN model. The accuracy of the final results could be affected if little or no correlation exists between the variables used as input values in the ANN model and the energy consumption of the chiller. Therefore, the correlation between the energy consumption of the chiller and the Spearman rank-order correlation coefficient for each variable was analyzed. The Spearman rank-order correlation coefficient had a value between −1.0 and +1.0, similar to other correlation coefficients.

Table 3 presents the Spearman rank-order correlation coefficient for each variable used for the ANN analysis of the proposed chiller's energy consumption. Nine variables were originally considered: Chilled water flow rate, cooling water temperature, outside dry-bulb temperature, outside wet-bulb temperature, dew-point temperature, outside relative humidity, hours, type of day, and supply chilled water temperature. The correlation coefficients indicate that the outside dry-bulb temperature, outside wet-bulb temperature, and outside relative humidity are the main parameters. The negative correlation coefficient of supply chilled water temperature results in the increase of chiller energy consumption as the water temperature decreased. Table 3 shows the correlation between variables and chiller energy consumption.

Table 3. Correlation between variables and chiller energy consumption.

Variables	Chilled Water Flow Rate (kg/s)	Cooling Water Temp. (°C)	Outside Dry-Bulb Temp. (°C)	Outside Wet-Bulb Temp. (°C)	Dew-Point Temp. (°C)	Outside Relative Humidity (%)	Hour	Type of Day	Supply Chilled Water Temp. (°C)
Spearman correlation coefficient	0.99	0.90	0.89	0.88	0.83	0.16	0.06	0.04	−0.67

2.3. Development of a Predictive Model of Energy Consumption of the Chiller Using the ANN Model

An ANN model is a network created by connecting nodes. It processes learning based on the weight of the nodes between the input and target values and outputs the result. An ANN model consists of an input layer, hidden layer, and output layer. An input value for training is derived and the input signal is transmitted to the next node in the input layer. The hidden layer is connected to all the nodes in the input layer, receives the input signal, and performs the neural network operation through the connection of the hidden layer nodes. Then, the output layer calculates the final result through the operation value of the hidden layer. For this study, the input values for the input layer were selected after making a list in the ANN to find the most accurate model to predict the energy consumption of the chiller. Figure 2 presents a schematic diagram of the ANN model used in this study as derived using MATLAB (version R2018a). A feed-forward automatic nonlinear NARX (nonlinear autoregressive network with exogenous) method, which uses measured values as inputs to dynamic neural networks, was employed to improve the predictive performance of the model. The NARX method is preferred to predict a time series dataset [20,21]. Figure 2 shows the schematic diagram of the predictive model of chiller energy consumption using the ANN model.

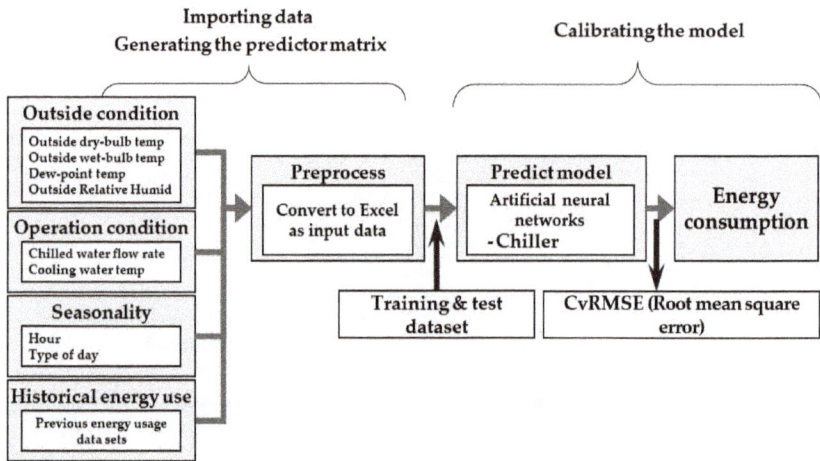

Figure 2. Schematic diagram of the predictive model of chiller energy consumption using the artificial neuron network (ANN) model.

Table 4 summarizes the input and output settings for the predictive model. The accuracy of the predicted results for chiller energy consumption according to the input setting was analyzed and tested to find the optimal conditions for the proposed ANN model. To do that, three input variables were considered: The number of input data points, the number of neurons and their size, and the amount (proportion) of the training data. By applying these input variables, the accuracy of the predicted results could be compared. Eight input values were used instead of the original nine, because the supply chilled water temperature had a negative correlation coefficient among the data generated. The eight remaining input values were added in order of their correlation coefficient from highest to lowest, and the accuracy of the results was compared according to the number of input values, which correlated to eight 'cases' for this study.

Table 4. Input/output conditions for chiller energy prediction model.

Input Data	Chilled water flow rate (kg/s)
	Cooling water temperature (°C)
	Outside dry-bulb temperature (°C)
	Outside wet-bulb temperature (°C)
	Dew-point temperature (°C)
	Outside relative humidity (%)
	Hour (h)
	Type of day (weekdays, weekend)
Number of Neurons	2–20
Proportion of Training Data	50%–90% (of 3761 data sets)
Predicted Target Y(t)	Chiller energy consumption (kWh)

For the training data, the amount of the data shows that the accuracy of the predicted results varied between 50% and 90% of the 3761 datasets that corresponded to the cooling period (warmest seasons) from May to September. Those data were generated based on the reference building model discussed in Section 2.1. The number of neurons is also one of the most important variables in a neural network. In this study, the accuracy of the results and computation speed were compared based on a range from two to twenty neurons. Table 4 shows the input/output conditions for the chiller energy prediction model.

ASHRAE (American Society of Heating, Refrigeration, and Air Conditioning Engineers) Guideline 14, Measurement of Energy and Demand Savings [22], was used to confirm that the test results were reliable when the tolerance limits were within the range of specified tolerances, as shown in Table 5. The accuracy and reproducibility of the predictive model were verified through the CvRMSE of the results obtained from 10 runs per condition. The chiller energy usage was predicted using the conditions with the highest accuracy. Table 5 shows the acceptable calibration tolerances.

Table 5. Acceptable calibration tolerances.

Calibration Type	Index	Acceptable Value *
Monthly	MBE	±5%
	CvRMSE	15%
Hourly	MBE	±10%
	CvRMSE	30%

Note: MBE is mean bias error; CvRMSE is the coefficient of variance of the root mean squared error. * Lower values indicate better calibration.

MBE and CvRMSE are defined by Equations (1) and (2):

$$\text{MBE} = \left\{ \left[\sum_{i=1}^{n} (y_i - \hat{y}_i) \right] / [(n-p) \times \overline{y}] \right\} \times 100, \tag{1}$$

$$\text{CvRMSE} = 100 \times \left[\sum (y_i - \hat{y}_i)^2 / (n-p) \right]^{1/2} / \overline{y}, \tag{2}$$

where n is the number of data points, p is the number of parameters, y_i is the utility data used for calibration, $\hat{y}_i$ is the simulation predicted data, and $\overline{y}$ is the arithmetic mean of the sample of n observations.

3. Results and Discussion

The accuracy of the predictions of the chiller's energy consumption was based on the results for the number of input variables, amount (proportion) of the data, and number of neurons during both the training and testing periods. The following sections present and discuss these results.

3.1. Effect of the Number of Input Variables for the Training Period and Testing Period

The accuracy of the predicted results was investigated according to the number of input variables. The input variables were added sequentially one by one starting from the chilled water flow rate with the highest correlation coefficient, as summarized in Table 6. The amount of training data was fixed at 50% of the total dataset, and the number of neurons was fixed at 20. Table 6 shows the conditions for the input variables.

Figure 3 shows the CvRMSE of the predicted energy consumption of the chiller for each case (number of inputs) during the training period. For most of the cases, the predicted values did not exceed the ASHRAE Guideline standard of 30% at a fixed training period of 50%. In Case 1 with one input parameter (chilled water flow rate), the repeatability was good with a standard deviation of 0.3. The results were comparable with Case 4 and Case 5, even when the number of input variables was only one. Case 2 and Case 3 were 57.7% and 31.3%, respectively, which exceed the limit of 30%. If the number of input variables is few, this factor will affect the reproducibility of the predicted results. When the number of input variables was 7 (min. 17.7%, max. 21.2%, and mean 19.8%) or 8 (min. 17.5%, max. 21.1%, and mean 19.5%), the predicted results were more accurate than for the other conditions. Figure 3 shows the accuracy according to the number of input variables for the training period.

Table 6. Conditions for the input variables.

Number of Inputs	Chilled Water Flow Rate (kg/s)	Cooling Water Temp. (°C)	Outside Dry-Bulb Temp. (°C)	Outside Wet-Bulb Temp. (°C)	Dew-Point Temp. (°C)	Outside Relative Humidity (%)	Hour	Type of Day
				Possible Input Variables				
Spearman correlation coefficient	0.99	0.90	0.89	0.88	0.83	0.16	0.06	0.04
Case 1	○							
Case 2	○	○						
Case 3	○	○	○					
Case 4	○	○	○	○				
Case 5	○	○	○	○	○			
Case 6	○	○	○	○	○	○		
Case 7	○	○	○	○	○	○	○	
Case 8	○	○	○	○	○	○	○	○

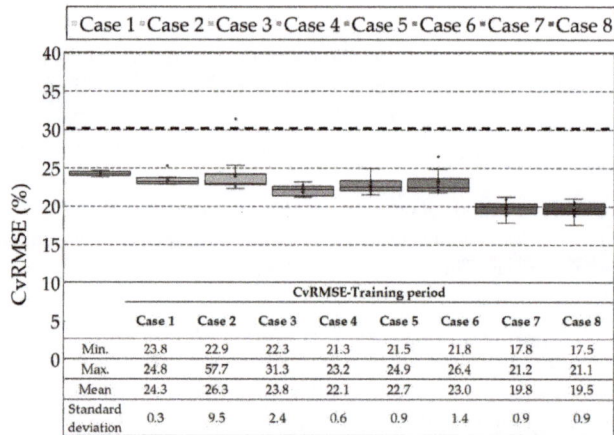

Figure 3. Accuracy according to the number of input variables for the training period.

Figure 4 shows the CvRMSE of the predicted energy consumption of the chiller during the testing period for each case (number of inputs). For most cases, the CvRMSE is in the range of 30% to 60%. The accuracy and reproducibility were shown to have decreased compared to the CvRMSE results during the training period. However, the accuracy and reproducibility of the predicted results gradually improved as the number of input variables increased. When the number of inputs was more than 7, the CvRMSE was an average of less than 30%, which is the ASHRAE standard. Case 8 shows the best results with a min. of 19.4%, max. of 30.2%, mean of 22.8%, and standard deviation of 3.0. The predicted values for energy usage during the training interval and testing interval according to the variation in input variables confirmed that the prediction accuracy improved as the number of variables increased. Even though variables such as outside relative humidity, hour, and type of day were not closely correlated with chiller energy usage, but those variables also helped to improve accuracy. Figure 4 shows the accuracy according to the number of input variables for the testing period.

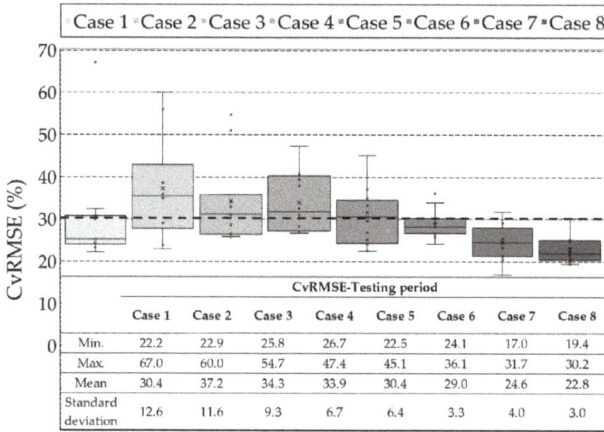

Figure 4. Accuracy according to the number of input variables for the testing period.

3.2. Effect of Amount of Data For Training Period and Testing Period

The accuracy of the predicted energy consumption was evaluated also by varying the amount of training data from 50% to 90%, while the number of input variables and the number of neurons were fixed at 8 and 20, respectively. Table 7 shows the conditions for training and testing data size.

Table 7. Conditions for training and testing data size.

	Case 9		Case 10		Case 11		Case 12		Case 13	
	Training period	Testing period	Training period	Testing period	Training period	Testing period	Training period	Testing period	Training period	Testing period
Data size (%)	50	50	60	40	70	30	80	20	90	10

Figure 5 shows the accuracy for energy consumption with respect to the amount of the training dataset. For the training period, the results confirmed that the accuracy improved as the proportion of data used for prediction was increased. The reproducibility and accuracy of the predictions were excellent, with a standard deviation less than 1 for all cases. Figure 5 shows the accuracy according to changes in the proportion of training data for the training period.

Figure 6 presents the prediction accuracy for energy consumption with respect to the proportion of the testing dataset in terms of the percentage used during the testing period. The best results were obtained with the CvRMSE average of 18.2% and standard deviation of 1.1 when 60% and 40% of the training data were used, respectively. The accuracy and reproducibility of the prediction decreased as the amount of data used during the testing period was decreased. Since ANNs provide a technique for performing predictions through data learning, the amount of datasets used for learning has a great influence on the accuracy of the predictions. Therefore, the accuracy varied according to the proportion of overall data used respectively in the training period and testing period. In this study, the predicted results indicated that an optimal method to obtain accurate prediction results was to proportion the data to be 60% for the training period and 40% for the testing period. Figure 6 shows the accuracy according to changes in proportion of the testing data for the testing period.

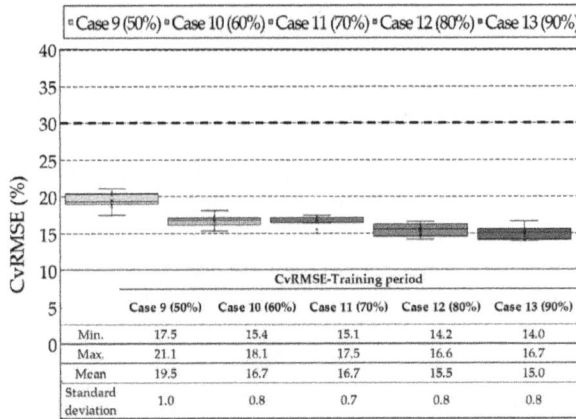

Figure 5. Accuracy according to changes in the proportion of training data for the training period.

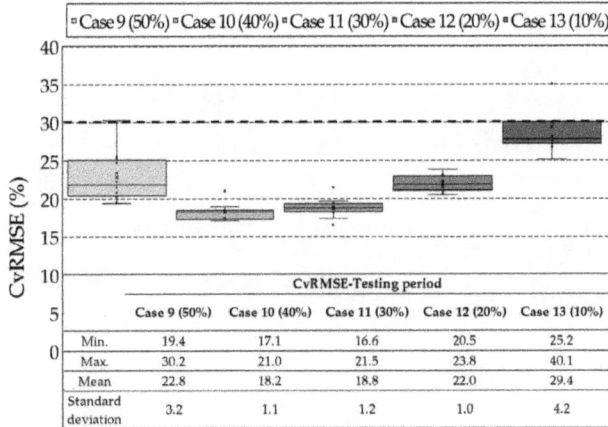

Figure 6. Accuracy according to the changes in proportion of the testing data for the testing period.

3.3. Effect of Number of Neurons for Training Period and Testing Period

In this study, the number of neurons was varied from 2 to 20, and the number of input variables was fixed at 8 with 60% training data. Figure 7 shows that the average CvRMSE was less than 20% accurate in every case except Case 14 (with two neurons, referred to as N2). These results indicate that increasing the number of neurons to a certain level improves the accuracy of the predicted results; however, no significant change was evident after more than 12 neurons were employed. Figure 7 shows the accuracy according to number of neurons used in the training period.

For the testing period, as shown in Figure 8, the mean value of the CvRMSE was also less than 20% accurate except for Case 14 (N2). No significant difference was evident in all cases, but when the number of neurons was 12 (N12), the best predicted result was obtained with the mean of 17.4% and standard deviation of 0.7. Based on these results, the use of 12 neurons showed the best accuracy for both the training period and testing period in this study. No significant effect on accuracy with respect to the number of neurons was evident because the optimized number of input variables and amount of training data were used. Since an increase in the number of neurons could delay the execution time for ANN algorithms, the number of neurons should be considered carefully after selecting the number of input variables and proportioning the amount of training data. Figure 8 shows the accuracy according to number of neurons for the testing period.

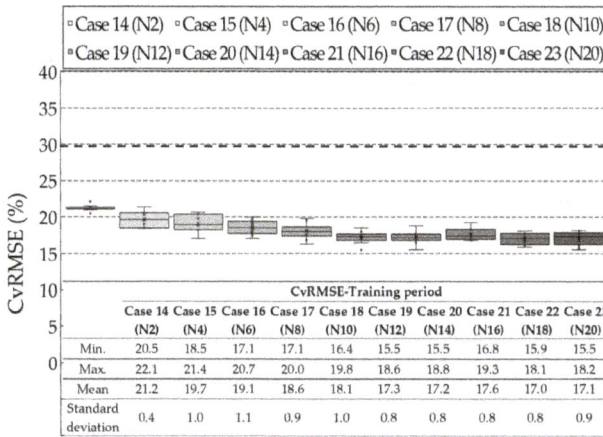

Figure 7. Accuracy according to the number of neurons used in the training period.

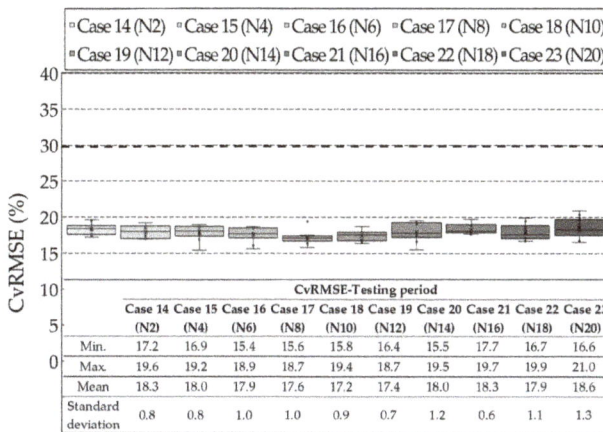

Figure 8. Accuracy according to the number of neurons for the testing period.

4. Prediction of Chiller Energy Consumption

The accuracy of the prediction of the energy consumption of the chiller was evaluated with respect to the various input conditions for the proposed ANN model. The condition that led to the highest level of accuracy was composed of eight input variables, 60% training data, and 12 neurons. The energy consumption of the chiller based on these derived optimal conditions was computed and compared with the data generated through the reference building in accordance with the large-scale office building used in DOE guidelines.

Figure 9 shows the energy consumption of the chiller system during warm/hot weather. The prediction period was from May to September, which comprises the seasons when cooling is most needed. The training period (60%) was selected as May to July and the testing period (40% of the dataset) was from August to September. Figure 9a presents a comparison of the monthly energy usage computed during the training period (from May to July) and the usage generated for the reference building. The error was 0.3%–2.4%. The error for August was 3.0% (predicted value of 127.7 MW versus the generated value of 131.6 MW). The error for September was 1.0% (predicted value of 88.3 MW versus the generated value of 87.5 MW). Figure 9b shows the total energy consumption prediction for the chiller with the error of 0.9% (predicted value of 488.1 MW versus the generated value of 492.7 MW) for the entire range of the cooling period (May to September). The predicted MW values closely matched the actual MW values. Figure 9a shows the monthly energy consumption and Figure 9b shows the total energy consumption.

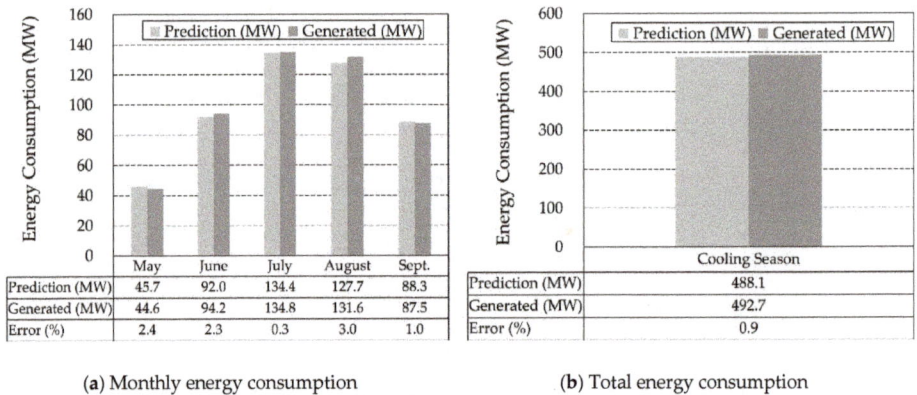

	May	June	July	August	Sept.
Prediction (MW)	45.7	92.0	134.4	127.7	88.3
Generated (MW)	44.6	94.2	134.8	131.6	87.5
Error (%)	2.4	2.3	0.3	3.0	1.0

	Cooling Season
Prediction (MW)	488.1
Generated (MW)	492.7
Error (%)	0.9

(a) Monthly energy consumption   (b) Total energy consumption

Figure 9. Prediction of chiller energy consumption (monthly for five months).

5. Conclusions

This study was conducted to find optimal conditions for a chiller in a centralized HVAC system by using an ANN algorithm. The developed chiller energy consumption model was evaluated for accuracy in terms of the following input parameters: Input conditions, number of input variables, amount of training data, and number of neurons. The limited findings were as follows.

With regard to optimizing the input variables, the prediction accuracy was secured in this study by increasing the number of input variables even if the correlation with the output value is low. With eight input variables, the CvRMSE reflected the highest accuracy of 19.5% and standard deviation of 0.9 in the training period, and the CvRMSE of 22.8% and standard deviation of 3.0 in the testing period.

With regard to optimizing the amount of training data, the prediction accuracy was similarly secured by increasing the percentage of the training data. However, increasing the training data means decreasing the testing data. The study results confirmed that prediction accuracy decreased gradually when the amount of data was decreased.

With regard to optimizing the number of neurons, when the number of input variables and amount of training data were fixed as per the previously verified conditions, no significant change in accuracy was found in terms of the number of neurons.

In order to obtain highly accurate predictions, various parameters such as conditions and number of input variables, sufficient available data, and the appropriate proportion of training versus testing data must be considered. In this study, by estimating the chiller energy usage based on eight input variables, 60% training data and 40% testing data, and 12 neurons, the predicted monthly energy consumption could be compared to the actual energy consumption generated by the DOE reference building. The comparison results indicated high prediction accuracy for the proposed chiller model with an error of only 0.9% of the total energy usage, which means that the proposed chiller was 99.1% accuracy.

For broadening the research, a deep learning model with more hidden layers and various cross validation method will be needed for the future works.

Energies **2019**, *12*, 2860

Author Contributions: J.-H.K. contributed to the project idea development and experimental design, and N.-C.S. performed the data analysis. W.C. reviewed the final manuscript and contributed to the results discussion and conclusions.

Funding: This work is supported by the Korea Agency for Infrastructure Technology Advancement(KAIA) grant funded by the Ministry of Land, Infrastructure and Transport (19AUDP-B099702-05).

Conflicts of Interest: The authors declare that there is no conflict of interests regarding the publication of this article.

References

1. Nassif, N. Modeling and Optimization of HVAC systems using Artificial Neural Network and Genetic Algorithm. *Int. J. Build. Simul.* **2014**, *7*, 237–245. [CrossRef]
2. Chou, J.S.; Bui, D.K. Modeling heating and cooling loads by artificial intelligence for energy-efficient building design. *Energy Build.* **2014**, *82*, 437–446. [CrossRef]
3. Deb, C.; Eang, L.S.; Yang, J.; Santamouris, M.; Santamouris, M. Forecasting diurnal cooling energy load for institutional buildings using Artificial Neural Networks. *Energy Build.* **2016**, *121*, 284–297. [CrossRef]
4. Cao, J.; Chen, C.; Hu, M.; Leung, M.K.H.; Pei, G. Numerical analysis of a novel household refrigerator with controllable loop thermosyphons. *Int. J. Refrig.* **2019**, *104*, 134–143. [CrossRef]
5. Cao, J.Y.; Chen, C.X.; Gao, G.T.; Yang, H.L.; Su, Y.H.; Bottarelli, M.; Cannistraro, M.; Pei, G. Preliminary evaluation of the energy-saving behavior of a novel household refrigerator. *J. Renew. Sustain. Energy* **2019**, *11*, 015102. [CrossRef]
6. Cannistraro, M.; Mainardi, E.; Bottarelli, M. Testing a dual-source heat pump. *Math. Model. Eng. Probl.* **2018**, *5*, 205–210. [CrossRef]
7. Jang, J.H.; Leigh, S.B. A Prediction of Optimal Heating Timing based on Artificial Neural Network by utilizing BEMS data. *Spring Conf. AIK* **2017**, *37*, 563–564. (In Korean)
8. Jeong, J.H.; Chae, Y.T. A Study on selection of Machine Learning types for Building Energy Consumption Forecasting. *2017 Summer Annu. Conf. SAREK* **2017**, 93–94. (In Korean). Available online: http://www.sarek.or.kr/html/sub06_02b.jsp?yearmonth=2017&quarter=s (accessed on 25 July 2019).
9. Jeong, J.H.; Chae, Y.T. Assessment of Input Variable Importance and Machine Learning Model Selection for Improving Short Term Load Forecasting on Different Building Types. *J. Korean Inst. Arch. Sustain. Environ. Build. Syst.* **2017**, *11*, 586–598.
10. Jeon, B.K.; Kim, E.J. Short-Term Load Prediction Using Artificial Neural Network Models. *Korean J. Air Cond. Refrig. Eng.* **2017**, *29*, 97–503. (In Korean)
11. Park, B.R.; Choi, E.; Moon, J.W. Performance tests on the ANN model prediction accuracy for cooling load of buildings during the setback period. *KIEAE J.* **2017**, *17*, 83–88. [CrossRef]
12. Ahmad, M.W.; Mourshed, M.; Rezgui, Y. Trees VS Neurons: Comparison between random forest and ANN for high-resolution prediction of building energy consumption. *Energy Build.* **2017**, *147*, 77–89. [CrossRef]
13. Seong, N.C.; Kim, J.H.; Choi, W.; Yoon, S.C.; Nassif, N. Development of Optimization Algorithms for Building Energy Model Using Artificial Neural Networks. *J. Korean Soc. Living Environ. Syst.* **2017**, *24*, 29–36. [CrossRef]
14. Seong, N.C.; Choi, W.C. Development of Predictive Fan Model using the Artificial Neural Network. *Autumn Annu. Conf. AIK.* **2017**, *37*, 604–607. (In Korean). Available online: https://www.aik.or.kr/html/page04_04_list.jsp?ncode=p002h (accessed on 25 July 2019).
15. Kim, J.H.; Seong, N.C.; Choi, W.C.; Choi, K.B. An Analysis of the Prediction Accuracy of HVAC Fan Energy Consumption According to Artificial Neural Network Variables. *J. Archit. Inst. Korea* **2018**, *34*, 73–79. (In Korean) [CrossRef]
16. Yu, F.W.; Chan, K. Improved energy management of chiller systems by multivariate and data envelopment analyses. *Appl. Energy* **2012**, *92*, 168–174. [CrossRef]
17. Thangavelu, S.R.; Myat, A.; Khambadkone, A. Energy optimization methodology of multi-chiller plant in commercial buildings. *Energy* **2017**, *123*, 64–76. [CrossRef]
18. Alonso, S.; Morán, A.; Prada, M.Á.; Reguera, P.; Fuertes, J.J.; Domínguez, M. A Data-Driven Approach for Enhancing the Efficiency in Chiller Plants: A Hospital Case Study. *Energies* **2019**, *12*, 827. [CrossRef]

Energies **2019**, *12*, 2860

19. Building Energy Codes Program. Available online: https://www.energycodes.gov/development/commercial/prototype_models. (accessed on 24 July 2019).
20. Boussaada, Z.; Curea, O.; Remaci, A.; Camblong, H.; Bellaaj, N.M. A Nonlinear Autoregressive Exogenous (NARX) Neural Network Model for the Prediction of the Daily Direct Solar Radiation. *Energies* **2018**, *11*, 620. [CrossRef]
21. Ruiz, L.G.B.; Cuéllar, M.P.; Calvo-Flores, M.D.; Jiménez, M.D.C.P. An Application of Non-Linear Autoregressive Neural Networks to Predict Energy Consumption in Public Buildings. *Energies* **2016**, *9*, 684. [CrossRef]
22. ASHRAE. Measurement of Energy and Demand Savings. *ASHRAE Guidel.* **2002**, *14*, 10–21.

energies

MDPI

Article

Enhancing the Heat Transfer in an Active Barocaloric Cooling System Using Ethylene-Glycol Based Nanofluids as Secondary Medium

Ciro Aprea [1], Adriana Greco [2], Angelo Maiorino [1] and Claudia Masselli [1,*]

[1] Department of Industrial Engineering, University of Salerno, Via Giovanni Paolo II 132, 84084 Fisciano (SA), Italy
[2] Department of Industrial Engineering, University of Naples Federico II, P.le Tecchio 80, 80125 Napoli, Italy
* Correspondence: cmasselli@unisa.it

Received: 16 July 2019; Accepted: 26 July 2019; Published: 28 July 2019

Abstract: Barocaloric cooling is classified as environmentally friendly because of the employment of solid-state materials as refrigerants. The reference and well-established processes are based on the active barocaloric regenerative refrigeration cycle, where the solid-state material acts both as refrigerant and regenerator; an auxiliary fluid (generally water of water/glycol mixtures) is used to transfer the heat fluxes with the final purpose of subtracting heat from the cold heat exchanger coupled with the cold cell. In this paper, we numerically investigate the effect on heat transfer of working with nanofluids as auxiliary fluids in an active barocaloric refrigerator operating with a vulcanizing rubber. The results reveal that, as a general trend, adding 10% of copper nanoparticles in the water/ethylene-glycol mixture carries to +30% as medium heat transfer enhancement.

Keywords: nanofluids; caloric cooling; barocaloric; solid-state cooling; acetoxy silicone rubber; Cu nanofluids; ethylene-glycol nanofluids; heat transfer

---

1. Introduction

The 21st century is characterized by a strong increase in energy demand. The World Bank Open Data assert that the worldwide per capita energy consumption rose from 1400 kWh in 1975 to 3200 kWh in our current days [1]. This hunger for energy impacts all sectors, among the most crucial of which are refrigeration and air-conditioning fields, to which more than 20% of the total world energy consumption is attributed. The need to propose alternative solutions to the now well-established cooling techniques has been a hot topic since the 1990s, alongside the first measures adopted by the Montreal Protocol [2], the Kyoto Protocol [3], and the following amendments [4], which had the aim of reducing environmental impact due to the refrigerant fluids used in vapor compression. The initial immediate measures adopted by the scientific community were devoted to replacing the most damaging fluids with substitute refrigerants [5–7]. Subsequently, there was a push to extend the research to alternative techniques that could completely change the ways of refrigerating and air conditioning [8–11]. One possible solution for supplying compensation for the use of vapor compression is represented by the Not-In-Kind cooling technologies [12] with specific emphasis on caloric refrigeration [13,14].

Caloric refrigeration embraces all the cooling techniques and was founded on a physical phenomenon according to which, due to an adiabatic change in the intensity of an external field, a variation of temperature (ΔT_{ad}) is detected in a solid-state material to which the field is applied [15,16]. The specificity of the field particularizes such effects classified as caloric; a magnetic field generates a magnetocaloric effect (MCE) [17–19], electric fields are associated with electrocaloric effects (ECE) [20–22], and mechanical fields provoke elastocaloric (eCE) [23] or barocaloric effects (BCE) [24], respectively, as consequences of stretching or of hydrostatic pressure application.

Caloric cooling is classified as environmentally friendly because it employs solid-state materials as refrigerants that do not directly impact global warming since they do not disperse in the atmosphere, as confirmed by a certain number of investigations [25–28] that asserted the eco-friendliness of all the techniques belonging to magneto- [29,30], electro-[31,32], elasto- [33,34], and baro-caloric [35] cooling.

The reference and well-established systems for caloric cooling are based on the active caloric regenerative refrigeration (ACR) cycle, a thermodynamic Brayton-based cycle in which the caloric solid-state material acts as both the refrigerant and the regenerator, thus recovering heat fluxes through the help of an auxiliary fluid that vehiculates them with the final purpose of subtracting heat from the cold heat exchanger (and therefore the cold environment) [36]. To improve the efficiency of a caloric cooler, the ACR system works with the optimal operative parameters (such as the geometry of the regenerator, the fluid velocity, and the frequency of the cycle) [37–39] but the bottleneck is employing materials due of high caloric effect in the temperature range toward the application is devoted to [40]. A very wide "chapter" is currently open in the research panorama regarding the efforts to realize suitable caloric effect materials. A huge number of studies have been conducted over the last decades, and several promising materials have been identified for each specific caloric technique [41–43].

Magnetocaloric is the most mature solid-state cooling technique [44,45], as it was the first to be investigated; however, there are still inherent limitations and holdups in creating high-intensity magnetic fields by permanent magnet application [46,47]. More recently, electrocaloric and elastocaloric techniques have gained interest in the scientific community—certain objectives are being reached, and many prototypes have been presented and patented [48–50]. The barocaloric technique is the more embryonic caloric cooling technique. Initial studies have already identified promising barocaloric materials due to enhanced BCE that make it suitable for refrigeration and heat pump applications [51,52]. Vulcanizing natural rubbers and elastomers materials based on polydimethylsiloxane that exhibit giant barocaloric effects have gained certain interest [53,54].

On the one hand, important steps have been taken in researching caloric materials. However, on the other, much less attention has been paid to the secondary fluid of the active caloric system. Until now, the most common auxiliary fluids were water or water-glycol mixtures, according to the operating temperature ranges of the systems (above or below the zero Celsius point). Since the auxiliary fluid is responsible for the heat vehiculation in the regenerator from one side to another, optimizing solutions should be proposed for enhancing the heat transfer between the solid-state refrigerant and the fluid itself. One of these is constituted by nanofluids.

Nanofluids are classified as a new kind of heat transfer fluid; they are formed by common base fluids in which solid-state nanoparticles are dispersed due to high thermal conductivity. The nanoparticles, as the name suggests, present nanometric dimensions (< 100 nm) and could have metallic or non-metallic natures. The resulting fluid is a stable mixture of a base fluid with nanoparticles—called nanofluid—due to the enhanced thermal conductivity with respect to the starting base fluid [55]. In 1995, Choi [56] presented the scientific community with the concept of nanofluids, and the discovery was revolutionary. Over the years, nanofluids attracted many scientists from all over the world, who started to test them in wide-ranging applications [57–63]. Therefore, the use of nanofluids seems to be very promising in enhancing the thermal conductivity of a base fluid, but attention must be paid to the synthesis and the characterizations of them. Depending on the dimensions and the nature of the base fluid and the nanoparticles, the most appropriate methods for nanofluids synthesis must be chosen to avoid undesired effects, such as agglomeration, which takes place during the process of drying, storage, and transportation of nanoparticles. Agglomeration causes settlement and clogging of the channels, and the thermal conductivity of the nanofluids is consequently decreased. In some cases, to avoid potential agglomeration, specific processes of synthesis of the nanofluids could require high costs, and this could represent a drawback in their employment [64].

In regards to solid-state refrigeration, just two methods have been studied for the application of nanofluids in the caloric systems, and both were focused on magnetocaloric refrigeration [65,66]. Working with nanofluids in other caloric-effects-based cooling systems is still an unexplored field. In this paper, we introduce a numerical study conducted on an active barocaloric refrigerator operating with a vulcanizing rubber, which employed glycol-based nanofluids as auxiliary fluid. The tool employed for the investigation was a two-dimensional model of a parallel-plates active caloric regenerator, which was already validated experimentally in our previous investigations [16,67,68]. The analysis is focused mainly on the heat transfer performances of the regenerator.

2. Thermodynamic Cycle and the Tool of the Investigation

2.1. ABR Cycle: The Thermodynamic Cycle for Barocaloric Cooling

The active caloric regenerative refrigeration cycle applied to barocaloric cooling is defined as the ABR cycle (active barocaloric regenerative refrigeration cycle). As is visible in Figure 1, the ABR cycle is composed of four processes that are repeated cyclically in a refrigeration system:

A.   Adiabatic compression;
B.   Heat vehiculation from Cold Heat EXchanger (CHEX) to Hot Heat EXchanger (HHEX);
C.   Adiabatic decompression;
D.   Heat vehiculation from HHEX to CHEX.

The cycle is experimented upon by a barocaloric system formed by a regenerator containing the barocaloric refrigerant and crossed by a thermo-vectoral fluid called heat transfer fluid (HTF), which connects CHEX with HHEX. A pressure cell is used to change the operation pressure.

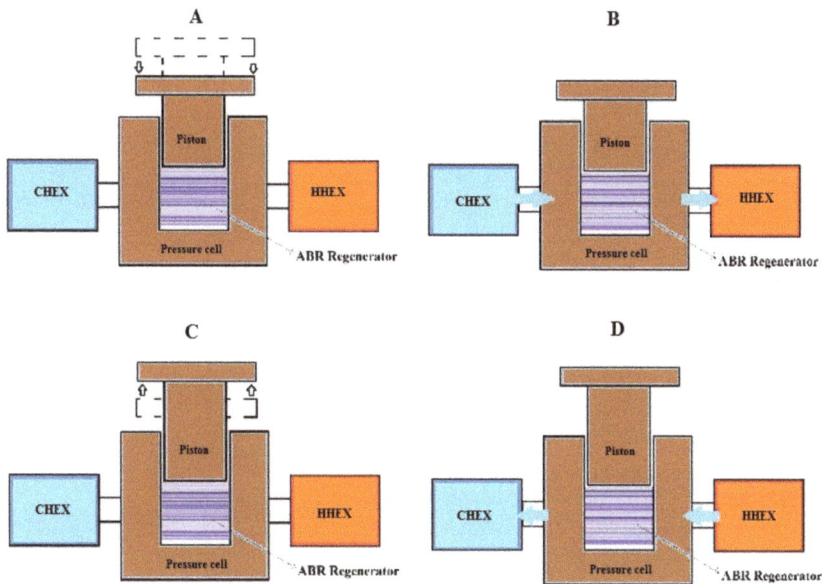

Figure 1. The four steps composing the active barocaloric regenerative (ABR) refrigerant cycle.

In the first step (A), the barocaloric regenerator is adiabatically compressed by means of the pressure cell; this causes the barocaloric effect to manifest with a rising temperature in the barocaloric material. After the compression, while keeping the pressure at maximum value, the HTF flows from the cold to the hot heat exchanger (B), thus the regenerator cools down, and the heat transferred to the fluid is released in the HHEX. When the fluid stops blowing, the adiabatic decompression begins (C), and pressure progressively decreases until reaching the minimum value. As a result, the regenerator cools further due to BCE. In the last step (D), the fluid crosses from HHEX to CHEX in the barocaloric regenerator, releasing heat to the latter and thus realizing the desired effect with the heat removed from CHEX.

2.2. Numerical Model

A two-dimensional (2-D) model of the parallel-plate matrix of the barocaloric material with stacking channels for HTF flow was employed to perform the investigation introduced in this paper.

Figure 2 shows the geometry of the model. The HTF flows along the x-axis in both directions. On the left side, by means of a first-type boundary condition, the presence of CHEX (with T_C as the set point temperature) is modeled; the same is done on the right side, where T_H is the first-type condition associated with HHEX.

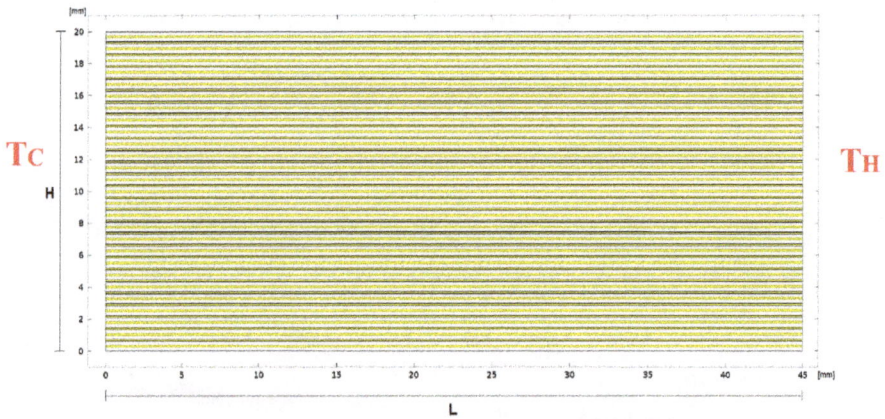

Figure 2. The geometry of the parallel-plates barocaloric regenerator.

The mathematical system that stems from forcing the ABR cycle on the geometry is composed of the following equations:

$$\begin{cases} \frac{\partial u}{\partial x} + \frac{\partial v}{\partial y} = 0 \\ \frac{\partial u}{\partial t} + u\frac{\partial u}{\partial x} + v\frac{\partial u}{\partial y} = -\frac{1}{\rho_{nf}}\frac{\partial p}{\partial x} + v\left(\frac{\partial^2 u}{\partial x^2} + \frac{\partial^2 u}{\partial y^2}\right) \\ \frac{\partial v}{\partial t} + u\frac{\partial v}{\partial x} + v\frac{\partial v}{\partial y} = -\frac{1}{\rho_{nf}}\frac{\partial p}{\partial y} + v\left(\frac{\partial^2 v}{\partial x^2} + \frac{\partial^2 v}{\partial y^2}\right) \\ \frac{\partial T_{nf}}{\partial t} + u\frac{\partial T_{nf}}{\partial x} + v\frac{\partial T_{nf}}{\partial y} = \frac{k_{nf}}{\rho_{nf}C_{nf}}\left(\frac{\partial^2 T_{nf}}{\partial x^2} + \frac{\partial^2 T_{nf}}{\partial y^2}\right) \\ \frac{\partial T_s}{\partial t} = \frac{k_s}{\rho_s C_s}\left(\frac{\partial^2 T_s}{\partial x^2} + \frac{\partial^2 T_s}{\partial y^2}\right) \end{cases} \tag{1}$$

$$\begin{cases} \rho_{nf}C_{nf}\frac{\partial T_{nf}}{\partial t} = k_{nf}\left(\frac{\partial^2 T_{nf}}{\partial x^2} + \frac{\partial^2 T_{nf}}{\partial y^2}\right) \\ \rho_s C_s\frac{\partial T_s}{\partial t} = k_s\left(\frac{\partial^2 T_s}{\partial x^2} + \frac{\partial^2 T_s}{\partial y^2}\right) + Q \end{cases} \tag{2}$$

$$Q = Q(p, T_S) = \frac{\rho_s\, C_s \Delta T_{ad}(p\,, T_s)}{\tau}, \tag{3}$$

The equations system (1), which defines the continuity of mass equation, the momentum equations of fluid (Navier-Stokes), and the fluid and the solid energy equations, is imposed during the fluid flow processes (steps B and D of the ABR, as shown in Figure 1). Laminar flow and incompressible fluid are assumed.

The equations system (2) acts during the phases of adiabatic compression and decompression. Because the fluid is not moving in these steps, it is composed only by the fluid and the solid energy equations. In the solid energy equation, the temperature change due to barocaloric effect is included by means of Q term, whose expression is explicated in Equation (3). Equation (3) practically converts the adiabatic temperature change due to BCE in a power density. The term depends on the working temperature of the barocaloric material and the applied pressure to the system. The related mathematical function for Q is evaluated through mathematical finder software following the elaboration and the manipulation of experimental data of specific heat and $\Delta T_{ad}(p, T_s)$, according to scientific literature. A complete description of all the steps we followed in the Q-terms building is reported in Aprea et al. [16]. Next to the construction and the manner of including Q in the field increasing/decreasing processes of the 2-D model, we detail how we accounted for the phenomenon of the thermal hysteresis that could become relevant for first order transition materials.

The model is solved through the finite element method (FEM). The ABR cycle is repeated cyclically many times until detecting that the regenerator has reached the steady-state condition. The latter must satisfy the cyclicality criterion in every point of the ABR regenerator:

$$\delta = max\{T(x, y, 0 + q\theta) - T(x, y, 4\tau + q\theta)\} < \bar{\varepsilon}, \tag{4}$$

The main peculiarity of this model is the extreme flexibility in making the caloric regenerator work with every caloric effect material (magneto-, electro-, elasto-, baro-), regardless of the specificity of the nature of the caloric effect exhibited. This extreme ductility constitutes a strongpoint of the tool, and it allows for the possibility of experimentally validating the model with just one of the four caloric effect-based prototypes. To this purpose, the model was introduced in our previous investigations [16,67] and validated with a prototype of a magnetocaloric cooler developed at University of Salerno. The validation, carried out both at zero load and with refrigerant load, showed fairly good agreement with experimental data [16,67,69].

3. Materials Employed in the Investigation

3.1. The Solid-State Barocaloric Refrigerant

The choice of the solid-state refrigerant to be employed in the active barocaloric system was done according to the requirements underlined by Aprea et al. [16] that a material must satisfy to be considered suitable for the specific caloric application.

We ultimately chose a barocaloric refrigerant with natural materials and elastomeric properties: acetoxy silicone rubber (ASR) [70]. Polydimethylsiloxane, additive, preserving agents and fillers are the components forming ASR, the latter exhibits a super-giant barocaloric effect with a peak of adiabatic change of temperature falling in correspondence of the crystalline–amorphous transitions with the consequence of polymer chains rearrangements. The adiabatic temperature change detected in acetoxy silicone rubber for decompression ($\Delta p = 0.390$ GPa) is shown in Figure 3.

The main thermodynamical and barocaloric features of ASR are reported in Table 1. A 41.1 K peak of adiabatic change of temperature occurs at 298 K. The latter data confers to ASR a relevant barocaloric effect in the temperature range where cooling and freezing applications are devoted to.

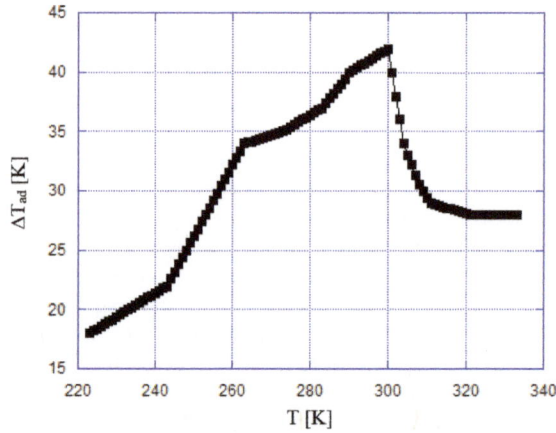

Figure 3. Temperature change due to barocaloric effects (BCE) and detected in acetoxy silicone rubber (ASR) during adiabatic decompression ($\Delta p = 0.390$ GPa).

Table 1. Thermal and physical features of acetoxy silicone rubber.

Material	T_{peak} [K]	Δp [GPa]	$\Delta T_{\text{ad,max}}$ [K]	Density [kg/m^3]	Thermal Conductivity [W/mK]
ASR	298	0.390	41.1	960	1.48

Following the methodology introduced in Section 2.2 about Q modeling, we built the mathematical expressions of Q terms for acetoxy silicone rubber. As an example, we report in Figure 4 the fitting of the Q term for the ASR representative of the adiabatic decompression process from 0.390 GPa to 0 GPa.

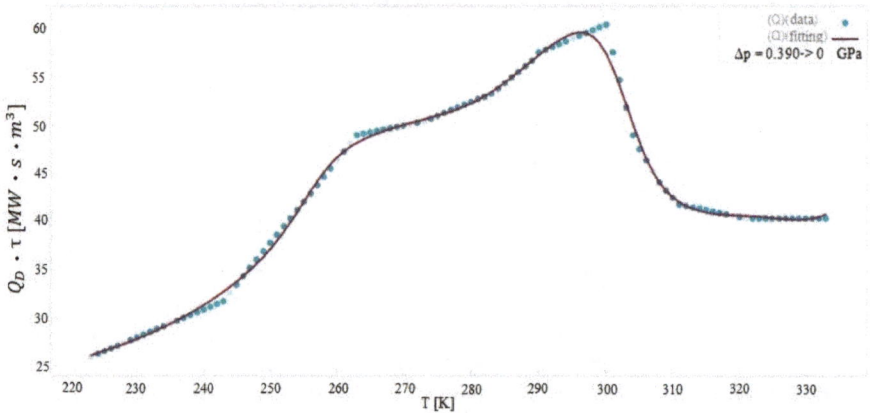

Figure 4. Q term constructing for ASR during adiabatic decompression process from 0.390 GPa to 0 GPa.

The resulting mathematical expression is:

$$Q_D \cdot \tau = 10^6 \left(1.33T + 0.000211 T^2 \sin(4.33 + \sin(-0.0636T)) sin(4.12 + \sin(0.0726T)) - 165 - 0.00216 T^2 \right), \quad (5)$$

3.2. Nanofluids as Heat Transfer Termvectorial Fluid

The role played by the secondary fluid is crucial for the heat vehiculation between the cold and the hot heat exchanger. Therefore, the heat transfer processes between the solid (parallel plates of barocaloric materials) and the fluid (auxiliary fluid) should be as efficient as possible so to allow the regenerator to work at a higher frequency in the ABR cycle. The optimization of the solid–fluid heat transfer process was not investigated deeply. Therefore, we proposed nanofluids as the solution to improve heat exchange, replacing the ordinary, common base fluids used until now (water or water-glycol mixtures).

The temperature range used in this investigation was one typical of refrigeration and freezing purposes; therefore, the starting base fluid was a 50% water − 50% ethylene glycol (EG) mixture, where the freezing point was 236 K. To improve the thermal conductivity of such an auxiliary fluid, Cu nanoparticles were dispersed. Thus, we chose to operate with metallic nanofluids because of the non-interaction of the metallic nature with the field generation for detecting the barocaloric effect (whose nature is mechanical).

As a result, the heat transfer fluids considered in the investigation and presented in this paper were Cu + 50% water − 50% EG nanofluids with variable concentrations of nanoparticles of nanometric dimensions (1÷10 nm). The thermophysical characteristics of the Cu nanoparticles are listed in Table 2.

Table 2. Thermal features of Cu.

Material	Specific Heat [J/kgK]	Density [kg/m^3]	Thermal Conductivity [W/mK]
Cu	383	8933	401

The considered thermophysical properties of the base fluid (50% water − 50% EG) were provided by the American Society of Heating, Refrigerating and Air-Conditioning Engineers (A. S. H. R. A. E.) as table functions of the temperature [71]. We elaborated such data in the resulting mathematical expression as follows:

$$\rho_{bf} = 1777 - 0.3482T, \tag{6}$$

$$C_{bf} = 2150 + 3.866T, \tag{7}$$

$$k_{bf} = 0.141 + 0.000769T, \tag{8}$$

$$\mu_{bf} = \frac{-0.6474}{229.5 - T} - 0.006321, \tag{9}$$

The thermophysical properties of the nanofluids were evaluated through the following mathematical correlations (available in open scientific literature) [72–74]:

$$\rho_{nf} = (1 - \varphi)\,\rho_{bf} + \varphi\,\rho_{np}, \tag{10}$$

$$\rho_{nf}C_{p,nf} = (1 - \varphi)\left(\rho C_p\right)_{bf} + \varphi\left(\rho C_p\right)_{np}, \tag{11}$$

$$\mu_{nf} = \frac{1}{(1 - \varphi)^{2.5}}\mu_{bf}, \tag{12}$$

$$k_{nf} = \frac{k_{np} + (n - 1)\,k_{bf} - (n - 1)\,\varphi\left(k_{bf} - k_{np}\right)}{k_{np} + (n - 1)\,k_{bf} + \varphi\left(k_{bf} - k_{np}\right)}, \tag{13}$$

where, supposing spherical nanoparticles, the empirical shape factor (n) is equal to 3.

4. Working Conditions of the Investigation

The model employed in this paper could reproduce the thermal-fluid dynamic behavior of an active caloric system for cooling purposes. In the investigation, the model was used to simulate the behavior of an ABR system working in refrigeration and freezing mode. The set-point temperatures of the cold and the hot heat exchangers were fixed, respectively, at $T_C = 255$ K and $T_H = 290$ K, and the frequency of the ABR cycle was 1.25 Hz. The system was tested at variable HTF velocities in the range $0.04 \div 0.2$ m/s. The Cu + 50% water − 50% EG nanofluid volume fraction also varied from 0% to 10%.

5. Results and Discussion

This paper mainly focused on the heat transfer between the solid and the fluid and on the investigation of possible solutions for its enhancement. Therefore, to this aim, we focus on one of the channels in which the HTF flows and there are two stacking parallel-plates of barocaloric material.

Figure 5 reports the mean temperature values of the boundaries of the channel (made of barocaloric material) and of the nanofluid flowing during the fourth step of the ABR cycle as a function of volume fraction and for a number of nanofluid velocities: (a) 0.04 m/s; (b) 0.06 m/s; (c) 0.10 m/s; (d) 0.20 m/s. From the figures, one can appreciate that the temperature difference between solid and nanofluid was reduced as nanoparticle volume fraction increased. This indicated that the heat exchange was improved following augmentation of φ. Furthermore, one can notice that increased fluid velocity caused a better exchange with the solid, and as a consequence, the fluid had a lower mean temperature, whereas the solid had a higher one. Figure 5c represents an exception, because the fluid temperature was slightly lower than the fluid with a velocity of 0.20 m/s (Figure 5d).

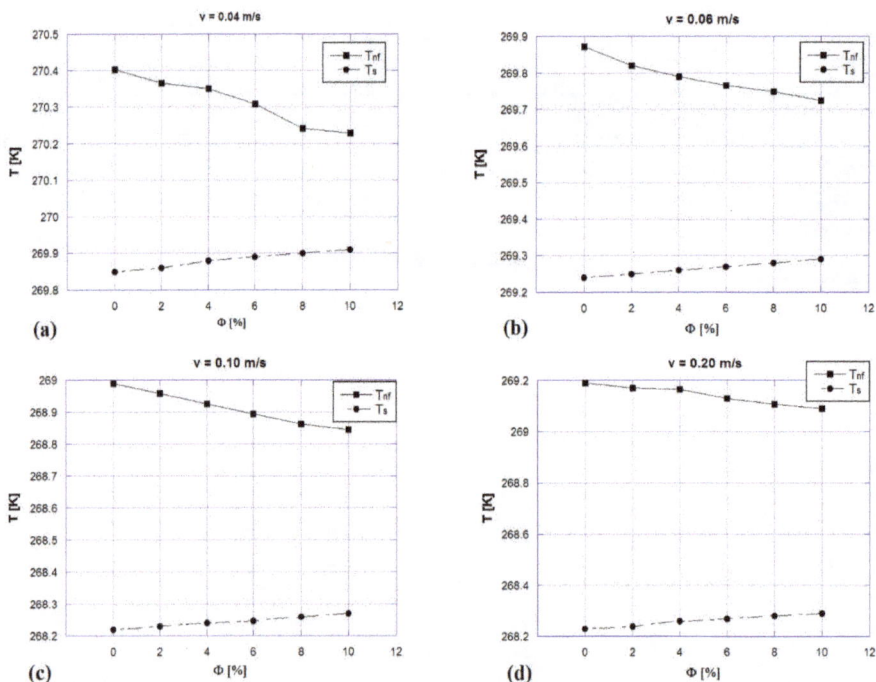

Figure 5. Solids and nanofluids temperatures averaged alongside one channel (x direction) and over time during the fourth step of the ABR cycle as a function of volume fraction and for the nanofluid velocities: (a) 0.04 m/s; (b) 0.06 m/s; (c) 0.10 m/s; (d) 0.20 m/s.

The most important indicator to evaluate the heat transfer performances between the solid and the fluid is represented by the convective heat transfer coefficient, defined as the proportionality coefficient between the convective heat flux at the solid-state boundaries of the channel and the temperature drop between solid and fluid. Therefore, in our investigation, we evaluated it as:

$$h = \frac{\dot{q}_c}{\Delta T} \tag{14}$$

The local convective heat flux was evaluated as:

$$\dot{q}_c(x,t) = k_{nf} \frac{\partial T_{nf}}{\partial y}, \tag{15}$$

whereas the temperature difference was evaluated as:

$$\Delta T = T_{nf} - T_s, \tag{16}$$

where both the partial derivative in Equation (15) and the temperature difference in Equation (16) were evaluated at the boundary between the fluid and the solid interface. To obtain the mean value of the heat transfer coefficient, both the local heat flux and the temperature difference were evaluated as mean values along the channel wall with respect to the fluid flow time.

Figure 6 reports the convective heat transfer coefficient as a function of nanofluid velocity parametrized for nanoparticle volume fraction. By fixing the velocity, one can appreciate the enhancement of the heat transfer coefficient with increasing nanoparticle volume fraction. A maximum h-enhancement of +34% was registered if the nanoparticles concentration increased from 0% up to 10% with $v = 0.06$ m/s, and +30% of heat transfer improvement for nanoparticles concentration increased from 0 to 10% was estimated as the mean value with respect to the nanofluid velocity. Moreover, one can notice that each curve plotted in Figure 6 had minimum points located at $v = 0.06$ m/s for low ϕ and at $v = 0.10$ m/s when the nanoparticles concentration increased. We found an explanation for this trend by analyzing each contribution of the nanofluid energy equation:

$$\rho_{nf} C_{nf} \left(\frac{\partial T_{nf}}{\partial t} + u \frac{\partial T_{nf}}{\partial x} + v \frac{\partial T_{nf}}{\partial y} \right) = k_{nf} \left(\frac{\partial^2 T_{nf}}{\partial x^2} + \frac{\partial^2 T_{nf}}{\partial y^2} \right) \tag{17}$$

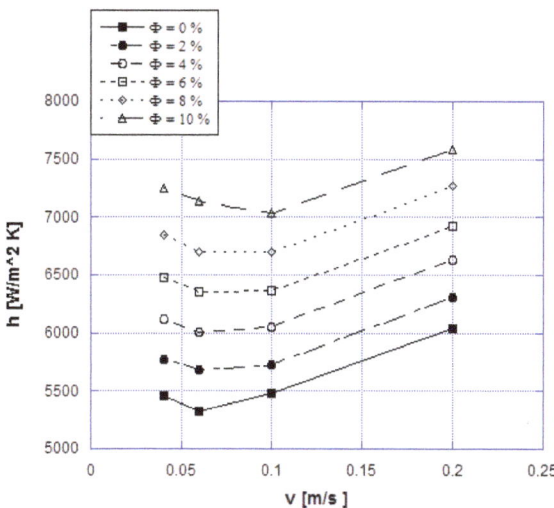

Figure 6. Convective heat transfer coefficient as a function of nanofluid velocity parametrized for nanoparticles volume fraction.

The left side represents the inertia and the convection terms, whereas, at the second member, there was the conduction term. Thus, the mean convective heat flux depended on fluid inertia, convection, and conduction. Convection and conduction promoted heat transfer, whereas inertia played a negative role.

At very low fluid velocities (0.04 m/s), inertia prevailed beyond conduction and convection, and this was detrimental for the heat exchange between fluid and solid. Consequently, the fluid temperature was higher, and this effect improved the fluid conductivity, increasing the conductivity heat exchange. By increasing the fluid velocity (0.06 m/s), the convection was slightly improved; therefore, the fluid temperature slightly decreased, consequently decreasing the conductive heat transfer. Therefore, the heat transfer coefficient slightly decreased. With a further increase of the fluid velocity (0.1 m/s), the convection term prevailed (rather than fluid inertia), and the heat transfer coefficient increased with fluid velocity. Therefore, at very low velocities, inertia played a dominant role in the heat transfer.

Indeed, taking as reference the convective heat transfer coefficients evaluated for the base fluid cases ($\phi = 0\%$), we estimated the corresponding h by means of the empirical correlations based on the dimensionless numbers of the forced convection (Reynolds and Prandtl numbers) and on the properties of the fluid itself at the reference temperature. The correlations used considered a fully developed and stationary laminar flow. Figure 7 reports a comparison between the convective heat transfer coefficients carried out from numerical simulations and the ones evaluated through empirical correlations for $\phi = 0\%$. Different trends could be observed, for example, the heat transfer coefficient from correlation was almost constant, whereas the numerical one first slightly decreased and then increased. Indeed, when the velocity increased over the threshold value, the convective term prevailed over the inertia, contributing to heat transfer enhancement. The correlations were based on steady state equations and therefore did not account for the inertia term. Furthermore, correlations always overpredicted heat transfer coefficients (from a minimum of +2.5% to a maximum of +16.5%) because they did not consider the detrimental effect of the inertia term. At the highest (0.2 m/s) fluid velocity, the difference between correlation and numerical model coefficient was at the minimum because the effect of inertia was lower.

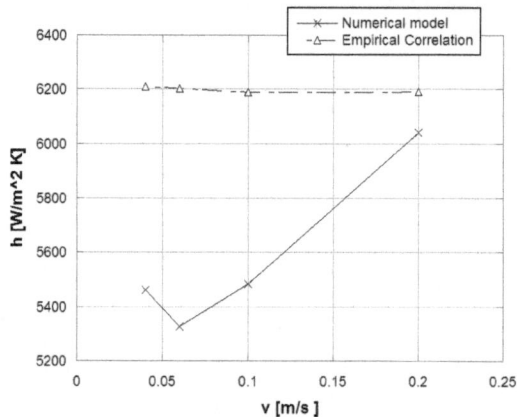

Figure 7. Comparison between the convective heat transfer coefficients carried out from the numerical model and the ones evaluated through empirical correlations.

To better analyze the effect of the heat transfer enhancement with nanofluids on the performance of the active barocaloric regenerator, we introduced an outstanding parameter as follows:

$$\Delta T_{ABR} = \frac{1}{\tau} \int_{\tau+n\theta}^{2\tau+n\theta} T_{nf}(L, y, t)dt - \frac{1}{\tau} \int_{3\tau+n\theta}^{4\tau+n\theta} T_{nf}(0, y, t)dt \,, \tag{18}$$

ΔT_{ABR} is a parameter used to evaluate the maximum temperature drop across the ABR since it measures the temperature difference between the nanofluid exiting the regenerator at the hot side ($x = L$) during the cold-to-hot flowing phase and at the cold side ($x = 0$) during the hot-to-cold flowing phase. To this hope, Figure 8 reports the trends of ΔT_{ABR} as a function of nanofluid velocity parametrized for nanoparticles volume fraction. As a general trend, it was possible to observe that there was a maximum of ΔT_{ABR}, located at $v = 0.06$ m/s for each volume fraction. At a fixed volume fraction, the ΔT_{ABR} first increased with fluid velocity, where it reached a maximum and then decreased. Indeed, by increasing the fluid velocity, the convective heat transfer coefficient also increased, as shown in Figure 6. Therefore, there was an optimal fluid velocity (around 0.06 m/s) corresponding to the point where the fluid could utilize the entire energy available from the barocaloric effect, and it was independent from the particles volume fraction. Further fluid flow perturbed the temperature profile of the ABR, reducing the temperature difference between the fluid and the regenerator. The latter quickly became overwhelmed by the fluid flow, and the efficiency of the heat transfer decreased. Figure 8 also shows that, at fixed fluid velocity, the ΔT_{ABR} increased by augmenting the particles volume fraction. Therefore, working with nanofluids had a positive influence on the convective barocaloric-refrigerant-auxiliary-fluid heat exchange. With increasing nanofluid velocity, the nanoparticle concentration had a lower influence on the augmentation ΔT_{ABR}, because the efficiency of the heat exchange decreased. Therefore, at $v = 0.2$ m/s, the points appeared closer to each other than had occurred at lower velocities.

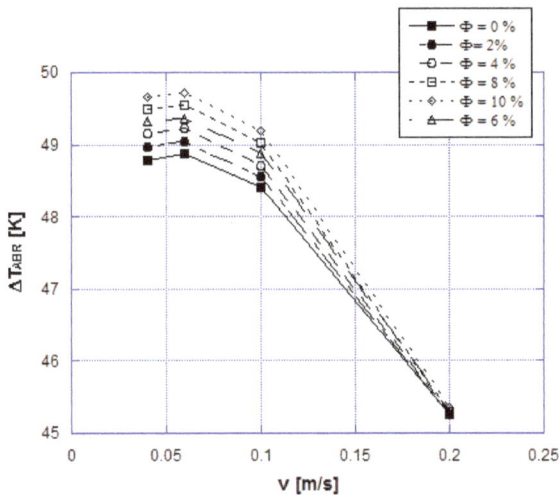

Figure 8. ΔT_{ABR} as a function of nanofluid velocity parametrized for nanoparticles volume fraction.

6. Conclusions

In this paper, the effects of working with nanofluids on heat transfer enhancement in an active barocaloric refrigerator were investigated through a two-dimensional finite element method. The ABR employed a barocaloric vulcanizing rubber as solid-state refrigerant and Cu + 50% ethylene glycol - 50% water nanofluids as auxiliary fluid. The nanoparticle volume fraction of the nanofluid varied from 0% to 10%, and the main indicators significative for evaluating the heat transfer were carried out. The results reveal that, as a general trend, the effect of adding 10% Cu nanoparticles in the water/ethylene-glycol mixture carries to +30% as a medium increment of heat transfer enhancement. Such an increment was detected with the ABR operating at different velocities and also influenced the energy parameters of the regenerator such as ΔT_{ABR}. Therefore, the addition of nanofluids in an active barocaloric refrigerator has many benefits.

Author Contributions: The following statements should be used "conceptualization, A.G., A.M. and C.M.; methodology, A.G.; software, C.M.; validation, A.G.; formal analysis, A.G. and C.M.; investigation, C.M.; resources, C.A. and A.M.; data curation, A.G., A.M. and C.M.; writing—original draft preparation, C.M.; writing—review and editing, A.G.; visualization, C.A. and A.M.; supervision, C.A., A.G. and A.M.; project administration, C.A. and A.G.; funding acquisition, C.A. and A.G."

Funding: This research received no external funding.

Conflicts of Interest: The authors declare no conflict of interest.

Nomenclature

Roman symbols

C	specific heat, $\text{J kg}^{-1}\text{K}^{-1}$
h	convective heat transfer coefficient, $\text{W m}^{-2}\text{K}^{-1}$
k	thermal conductivity, $\text{W m}^{-1}\text{K}^{-1}$
L	length of the regenerator in fluid flow direction, m
n	empirical shape factor
p	pressure, Pa
Q	power density associated to barocaloric effect, W m^{-3}
q	number of ABR cycles
$\dot{q}$	convective heat flux, W m^{-2}
T	temperature, K
t	time, s
u	longitudinal fluid velocity, m s^{-1}
v	orthogonal fluid velocity, m s^{-1}
x	longitudinal spatial coordinate, m
y	orthogonal spatial coordinate, m

Greek symbols

Δ	finite difference
∂	partial derivative
δ	infinitesimal difference
$\bar{\varepsilon}$	infinitesimal quantity
θ	period of the ABR cycle, s
μ	dynamic viscosity, Pa s
ν	cinematic viscosity, $\text{m}^2\text{ s}^{-1}$
ρ	density, kg m^{-3}
φ	volume fraction
τ	period of each step of the ABR cycle, s

Subscripts

ad	adiabatic
ABR	active barocaloric refrigerator
bf	base fluid
C	cold heat exchanger
c	convective
D	decompression
H	hot heat exchanger
nf	nanofluid
np	nanoparticles
s	solid

References

1. World Bank Open Data—Electric Power Consumption (kWh per Capita). Available online: https://data. worldbank.org/indicator/EG.USE.ELEC.KH.PC (accessed on 29 May 2019).
2. Protocol, Montreal. *Montreal Protocol on Substances that Deplete the Ozone Layer*; United Nation Environment Program (UN): New York, NY, USA, 1987.

3.   Protocol, Kyoto. *Kyoto Protocol to the United Nation Framework Convention on Climate Change*; United Nation Environment Program (UN): Kyoto, Japan, 1997.

4.   Heath, E.A. Amendment to the Montreal Protocol on Substances that Deplete the Ozone Layer (Kigali Amendment). *Int. Legal Mater.* **2017**, *56*, 193–205. [CrossRef]

5.   Greco, A.; Mastrullo, R.; Palombo, A. R407C as an alternative to R22 in vapour compression plant: An experimental study. *Int. J. Energy Res.* **1997**, *21*, 1087–1098. [CrossRef]

6.   Greco, A.; Vanoli, G.P. Flow boiling heat transfer with HFC mixtures in a smooth horizontal tube. Part II: Assessment of predictive methods. *Exp. Therm. Fluid Sci.* **2005**, *29*, 199–208. [CrossRef]

7.   Aprea, C.; Greco, A.; Maiorino, A. The substitution of R134a with R744: An exergetic analysis based on experimental data. *Int. J. Refrig.* **2013**, *36*, 2148–2159. [CrossRef]

8.   Brown, D.; Stout, T.; Dirks, J.; Fernandez, N. The Prospects of Alternatives to Vapor Compression Technology for Space Cooling and Food Refrigeration Applications. *Energy Eng.* **2012**, *109*, 7–20. [CrossRef]

9.   Goetzler, W.; Zogg, R.; Young, J.; Johnson, C. Alternatives to vapor-compression HVAC technology. *ASHRAE J.* **2014**, *56*, 12.

10. Bellos, E.; Tzivanidis, C. CO_2 Transcritical Refrigeration Cycle with Dedicated Subcooling: Mechanical Compression vs. Absorption Chiller. *Appl. Sci.* **2019**, *9*, 1605. [CrossRef]

11. Li, S.; Jeong, J.W. Energy Performance of Liquid Desiccant and Evaporative Cooling-Assisted 100% Outdoor Air Systems under Various Climatic Conditions. *Energies* **2018**, *11*, 1377. [CrossRef]

12. Brown, J.S.; Domanski, P.A. Review of alternative cooling technologies. *Appl. Therm. Eng.* **2014**, *64*, 252–262. [CrossRef]

13. Czernuszewicz, A.; Kaleta, J.; Lewandowski, D. Multicaloric effect: Toward a breakthrough in cooling technology. *Energy Convers. Manag.* **2018**, *178*, 335–342. [CrossRef]

14. Fähler, S.; Rößler, U.K.; Kastner, O.; Eckert, J.; Eggeler, G.; Emmerich, H.; Entel, P.; Müller, S.; Quandt, E.; Albe, K. Caloric Effects in Ferroic Materials: New Concepts for Cooling. *Adv. Eng. Mater.* **2012**, *14*, 10–19. [CrossRef]

15. Kitanovski, A.; Plaznik, U.; Tomc, U.; Poredoš, A. Present and future caloric refrigeration and heat-pump technologies. *Int. J. Refrig.* **2015**, *57*, 288–298. [CrossRef]

16. Aprea, C.; Greco, A.; Maiorino, A.; Masselli, C. Solid-state refrigeration: A comparison of the energy performances of caloric materials operating in an active caloric regenerator. *Energy* **2018**, *165*, 439–455. [CrossRef]

17. Pecharsky, V.K.; Gschneidner, K.A., Jr. Magnetocaloric effect and magnetic refrigeration. *J. Magn. Magn. Mater.* **1999**, *200*, 44–56. [CrossRef]

18. Ohnishi, T.; Soejima, K.; Yamashita, K.; Wada, H. Magnetocaloric Properties of (MnFeRu) 2 (PSi) as Magnetic Refrigerants near Room Temperature. *Magnetochemistry* **2017**, *3*, 6. [CrossRef]

19. Thang, N.; Dijk, N.; Brück, E. Tuneable giant magnetocaloric effect in (Mn, Fe) 2 (P, Si) materials by Co-B and Ni-B co-doping. *Materials* **2016**, *10*, 14. [CrossRef] [PubMed]

20. Scott, J.F. Electrocaloric materials. *Annu. Rev. Mater. Res.* **2011**, *41*, 229–240. [CrossRef]

21. Sun, X.; Huang, H.; Jafri, H.M.; Wang, J.; Wen, Y.; Dang, Z.M. Wide Electrocaloric Temperature Range Induced by Ferroelectric to Antiferroelectric Phase Transition. *Appl. Sci.* **2019**, *9*, 1672. [CrossRef]

22. Jiang, Z.Y.; Zheng, G.P.; Zheng, X.C.; Wang, H. Exceptionally high negative electro-caloric effects of poly (VDF–co–TrFE) based nanocomposites tuned by the geometries of barium titanate nanofillers. *Polymers* **2017**, *9*, 315. [CrossRef]

23. Wu, Y.; Ertekin, E.; Sehitoglu, H. Elastocaloric cooling capacity of shape memory alloys—Role of deformation temperatures, mechanical cycling, stress hysteresis and inhomogeneity of transformation. *Acta Mater.* **2017**, *135*, 158–176. [CrossRef]

24. Strässle, T.; Furrer, A. Cooling by adiabatic (DE) pressurization—The barocaloric effect. *High Press. Res.* **2000**, *17*, 325–333. [CrossRef]

25. Aprea, C.; Greco, A.; Maiorino, A.; Masselli, C. Magnetic refrigeration: An eco-friendly technology for the refrigeration at room temperature. *J. Phys. Conf. Ser.* **2015**, *655*, 012026. [CrossRef]

26. Saito, A.T.; Kobayashi, T.; Kaji, S.; Li, J.; Nakagome, H. Environmentally Friendly Magnetic Refrigeration Technology Using Ferromagnetic Gd Alloys. *Int. J. Environ. Sci. Dev.* **2016**, *7*, 316–320. [CrossRef]

27. Aprea, C.; Greco, A.; Maiorino, A.; Masselli, C. Electrocaloric refrigeration: An innovative, emerging, eco-friendly refrigeration technique. *J. Phys. Conf. Ser.* **2017**, *796*, 012019. [CrossRef]

28. Aprea, C.; Greco, A.; Maiorino, A.; Masselli, C. The environmental impact of solid-state materials working in an active caloric refrigerator compared to a vapor compression cooler. *Int. J. Heat Technol.* **2018**, *36*, 1155–1162. [CrossRef]

29. Kawanami, T.; Hirano, S.; Fumoto, K.; Hirasawa, S. Evaluation of fundamental performance on magnetocaloric cooling with active magnetic regenerator. *Appl. Therm. Eng.* **2011**, *31*, 1176–1183. [CrossRef]

30. Aprea, C.; Greco, A.; Maiorino, A.; Masselli, C. A comparison between rare earth and transition metals working as magnetic materials in an AMR refrigerator in the room temperature range. *Appl. Therm. Eng.* **2015**, *91*, 767–777. [CrossRef]

31. Valant, M. Electrocaloric materials for future solid-state refrigeration technologies. *Prog. Mater. Sci.* **2012**, *57*, 980–1009. [CrossRef]

32. Aprea, C.; Greco, A.; Maiorino, A.; Masselli, C. A comparison between different materials in an active electrocaloric regenerative cycle with a 2D numerical model. *Int. J. Refrig.* **2016**, *69*, 369–382. [CrossRef]

33. Wang, H.; Huang, H.; Xie, J. Effects of Strain Rate and Measuring Temperature on the Elastocaloric Cooling in a Columnar-Grained Cu71Al17. 5Mn11. 5 Shape Memory Alloy. *Metals* **2017**, *7*, 527. [CrossRef]

34. Tušek, J.; Engelbrecht, K.; Millán-Solsona, R.; Mañosa, L.; Vives, E.; Mikkelsen, L.P.; Pryds, N. The Elastocaloric Effect: A Way to Cool Efficiently. *Adv. Energy Mater.* **2015**, *5*, 1500361. [CrossRef]

35. Strässle, T.; Furrer, A.; Dönni, A.; Komatsubara, T. Barocaloric effect: The use of pressure for magnetic cooling in Ce 3 Pd 20 Ge 6. *J. Appl. Phys.* **2002**, *91*, 8543–8545. [CrossRef]

36. Plaznik, U.; Tušek, J.; Kitanovski, A.; Poredoš, A. Numerical and experimental analyses of different magnetic thermodynamic cycles with an active magnetic regenerator. *Appl. Therm. Eng.* **2013**, *59*, 52–59. [CrossRef]

37. Tušek, J.; Kitanovski, A.; Poredoš, A. Geometrical optimization of packed-bed and parallel-plate active magnetic regenerators. *Int. J. Refrig.* **2013**, *36*, 1456–1464. [CrossRef]

38. Aprea, C.; Greco, A.; Maiorino, A. An application of the artificial neural network to optimise the energy performances of a magnetic refrigerator. *Int. J. Refrig.* **2017**, *82*, 238–251. [CrossRef]

39. Teyber, R.; Trevizoli, P.V.; Christiaanse, T.V.; Govindappa, P.; Niknia, I.; Rowe, A. Semi-analytic AMR element model. *Appl. Therm. Eng.* **2018**, *128*, 1022–1029. [CrossRef]

40. Pecharsky, V.K.; Gschneidner, K.A., Jr. Advanced magnetocaloric materials: What does the future hold? *Int. J. Refrig.* **2006**, *29*, 1239–1249. [CrossRef]

41. Crossley, S.; Mathur, N.D.; Moya, X. New developments in caloric materials for cooling applications. *AIP Adv.* **2015**, *5*, 067153. [CrossRef]

42. Moya, X.; Kar-Narayan, S.; Mathur, N.D. Caloric materials near ferroic phase transitions. *Nat. Mater.* **2014**, *13*, 439–450. [CrossRef]

43. Liu, Y.; Scott, J.F.; Dkhil, B. Some strategies for improving caloric responses with ferroelectrics. *APL Mater.* **2016**, *4*, 064109. [CrossRef]

44. Yu, B.; Liu, M.; Egolf, P.W.; Kitanovski, A. A review of magnetic refrigerator and heat pump prototypes built before the year 2010. *Int. J. Refrig.* **2010**, *33*, 1029–1060. [CrossRef]

45. Sari, O.; Balli, M. From conventional to magnetic refrigerator technology. *Int. J. Refrig.* **2014**, *37*, 8–15. [CrossRef]

46. Franco, V.; Blazquez, J.S.; Amiano, A.C. Field dependence of the magnetocaloric effect in materials with a second order phase transition: A master curve for the magnetic entropy change. *Appl. Phys. Lett.* **2006**, *89*, 222512. [CrossRef]

47. Aprea, C.; Greco, A.; Maiorino, A. The use of the first and of the second order phase magnetic transition alloys for an AMR refrigerator at room temperature: A numerical analysis of the energy performances. *Energy Convers. Manag.* **2013**, *70*, 40–55. [CrossRef]

48. Correia, T.; Zhang, Q. *Electrocaloric Materials*; Springer: Berlin, Germany, 2014.

49. Qian, S.; Geng, Y.; Wang, Y.; Ling, J.; Hwang, Y.; Radermacher, R.; Takeuchi, I.; Cui, J. A review of elastocaloric cooling: Materials, cycles and system integrations. *Int. J. Refrig.* **2016**, *64*, 1–19. [CrossRef]

50. Greco, A.; Aprea, C.; Maiorino, A.; Masselli, C. A review of the state of the art of solid-state caloric cooling processes at room-temperature before 2019. *Int. J. Refrig.* **2019**. [CrossRef]

51. Lu, B.; Liu, J. Mechanocaloric materials for solid-state cooling. *Sci. Bull.* **2015**, *60*, 1638–1643. [CrossRef]

52. Aprea, C.; Greco, A.; Maiorino, A.; Masselli, C. A comparison between different materials with mechanocaloric effect. *Int. J. Heat Technol.* **2018**, *36*, 801–807. [CrossRef]

Energies **2019**, *12*, 2902

53. Bom, N.M.; Imamura, W.; Usuda, E.O.; Paixão, L.S.; Carvalho, A.M.G. Giant Barocaloric Effects in Natural Rubber: A Relevant Step toward Solid-State Cooling. *ACS Macro Lett.* **2017**, *7*, 31–36. [CrossRef]

54. Bom, N.M.; Usuda, E.O.; Guimarães, G.M.; Coelho, A.A.; Carvalho, A.M.G.; Carvalho, A. Note: Experimental setup for measuring the barocaloric effect in polymers: Application to natural rubber. *Rev. Sci. Instrum.* **2017**, *88*, 046103. [CrossRef]

55. Yu, W.; France, D.M.; Routbort, J.L.; Choi, S.U.S. Review and Comparison of Nanofluid Thermal Conductivity and Heat Transfer Enhancements. *Heat Transf. Eng.* **2008**, *29*, 432–460. [CrossRef]

56. Choi, S.U.S. Enhancing thermal conductivity of fluids with nanoparticles. *Dev. Appl. Non-Newton. Flows* **1995**, *66*, 99–105.

57. Mohamad, N.A.; Azis, N.; Jasni, J.; Kadir, A.; Abidin, M.Z.; Yunus, R.; Yaakub, Z. Impact of Fe_3O_4, CuO and Al_2O_3 on the AC Breakdown Voltage of Palm Oil and Coconut Oil in the Presence of CTAB. *Energies* **2019**, *12*, 1605. [CrossRef]

58. Irfan, S.A.; Shafie, A.; Yahya, N.; Zainuddin, N. Mathematical Modeling and Simulation of Nanoparticle-Assisted Enhanced Oil Recovery—A Review. *Energies* **2019**, *12*, 1575. [CrossRef]

59. Hussain, M.I.; Kim, J.H.; Kim, J.T. Nanofluid-Powered Dual-Fluid Photovoltaic/Thermal (PV/T) System: Comparative Numerical Study. *Energies* **2019**, *12*, 775. [CrossRef]

60. Fal, J.; Mahian, O.; Żyła, G. Nanofluids in the Service of High Voltage Transformers: Breakdown Properties of Transformer Oils with Nanoparticles, a Review. *Energies* **2018**, *11*, 2942. [CrossRef]

61. Kristiawan, B.; Santoso, B.; Wijayanta, A.; Aziz, M.; Miyazaki, T. Heat transfer enhancement of TiO2/water nanofluid at laminar and turbulent flows: A numerical approach for evaluating the effect of nanoparticle loadings. *Energies* **2018**, *11*, 1584. [CrossRef]

62. Sun, X.H.; Yan, H.; Massoudi, M.; Chen, Z.H.; Wu, W.T. Numerical simulation of nanofluid suspensions in a geothermal heat exchanger. *Energies* **2018**, *11*, 919. [CrossRef]

63. Bellos, E.; Tzivanidis, C. Optimization of a solar-driven trigeneration system with nanofluid-based parabolic trough collectors. *Energies* **2017**, *10*, 848. [CrossRef]

64. Kumar, S.A.; Meenakshi, K.S.; Narashimhan, B.R.V.; Srikanth, S.; Arthanareeswaran, G. Synthesis and characterization of copper nanofluid by a novel one-step method. *Mater. Chem. Phys.* **2009**, *113*, 57–62. [CrossRef]

65. Chiba, Y. Enhancements of thermal performances of an active magnetic refrigeration device based on nanofluids. *Mechanics* **2017**, *23*, 31–38. [CrossRef]

66. Mugica, I.; Roy, S.; Poncet, S.; Bouchard, J.; Nesreddine, H. Exergy analysis of a parallel-plate active magnetic regenerator with nanofluids. *Entropy* **2017**, *19*, 464. [CrossRef]

67. Aprea, C.; Cardillo, G.; Greco, A.; Maiorino, A.; Masselli, C. A comparison between experimental and 2D numerical results of a packed-bed active magnetic regenerator. *Appl. Therm. Eng.* **2015**, *90*, 376–383. [CrossRef]

68. Aprea, C.; Cardillo, G.; Greco, A.; Maiorino, A.; Masselli, C. A rotary permanent magnet magnetic refrigerator based on AMR cycle. *Appl. Therm. Eng.* **2016**, *101*, 699–703. [CrossRef]

69. Aprea, C.; Greco, A.; Maiorino, A.; Masselli, C. Energy performances and numerical investigation of solid-state magnetocaloric materials used as refrigerant in an active magnetic regenerator. *Therm. Sci. Eng. Prog.* **2018**, *6*, 370–379. [CrossRef]

70. Imamura, W.; Usuda, E.O.; Paixão, L.S.; Bom, N.M.; Gomes, A.M.; Carvalho, A.M.G. Supergiant barocaloric effects in acetoxy silicone rubber around room temperature. *arXiv* **2017**, arXiv:1710.01761.

71. Ashrae. *2009 ASHRAE Handbook—Fundamentals (SI Edition)*; Ashrae: Atlanta, GA, USA, 2009.

72. Bianco, V.; Vafai, K.; Manca, O.; Nardini, S. *Heat Transfer Enhancement with Nanofluids*; CRC Press: Boca Raton, FL, USA, 2015.

73. Brinkman, H.C. The viscosity of concentrated suspensions and solutions. *J. Chem. Phys.* **1952**, *20*, 571. [CrossRef]

74. Hamilton, R.L.; Crosser, O.K. Thermal conductivity of heterogeneous two-component systems. *Ind. Eng. Chem. Fund.* **1962**, *1*, 187–191. [CrossRef]

energies

MDPI

Article

Thermodynamic Evaluation of LiCl-H$_2$O and LiBr-H$_2$O Absorption Refrigeration Systems Based on a Novel Model and Algorithm

Jie Ren [1], Zuoqin Qian [1], Zhimin Yao [1,2,*], Nianzhong Gan [1] and Yujia Zhang [1]

[1] School of Energy and Power Engineering, Wuhan University of Technology, Wuhan 430063, China
[2] Faculty of Engineering and Information Technologies, University of Technology Sydney, Sydney, NSW 2007, Australia
* Correspondence: yaozm@whut.edu.cn

Received: 21 June 2019; Accepted: 1 August 2019; Published: 6 August 2019

Abstract: An absorption refrigeration system (ARS) is an alternative to the conventional mechanical compression system for cold production. This study developed a novel calculation model using the Matlab language for the thermodynamic analysis of ARS. It was found to be reliable in LiCl-H$_2$O and LiBr-H$_2$O ARS simulations and the parametric study was performed in detail. Moreover, two 50 kW water-cooled single effect absorption chillers were simply designed to analyze their off-design behaviors. The results indicate that LiCl-H$_2$O ARS had a higher coefficient of performance (*COP*) and exergetic efficiency, particularly in the lower generator or higher condenser temperature conditions, but it operated more restrictively due to crystallization. The off-design analyses revealed that the preponderant performance of LiCl-H$_2$O ARS was mainly due to its better solution properties because the temperature of each component was almost the same for both chillers in the operation. The optimum inlet temperature of hot water for LiCl-H$_2$O (83 °C) was lower than that of LiBr-H$_2$O (98 °C). The cooling water inlet temperature should be controlled within 41 °C, otherwise the performances are discounted heavily. The *COP* and cooling capacity could be improved by increasing the temperature of hot water or chilled water properly, contrary to the exergetic efficiency.

Keywords: absorption refrigeration system; thermodynamic analysis; calculation model; LiCl-H$_2$O; LiBr-H$_2$O; off-design behaviors

---

1. Introduction

As a promising technology to use low thermal potential energy and renewable energy, an absorption refrigeration system (ARS) is getting significant attention nowadays because it can be driven by waste heat [1], biomass [2], solar [3], geothermal energy [4], etc., instead of conventional compression-work-driven chillers. Furthermore, the ARS typically uses water as a refrigerant and the working fluids of system operation are environmentally friendly with zero global warming potential (GWP) and ozone depletion potential (ODP) [5].

In the ARS, the type of absorbent–refrigerant pair selected plays a major role in the performance. Sun et al. [6] has presented a complete review of working pairs of absorption cycles by dividing them into five general series according to the kinds of refrigerant. However, there are limited pairs commercially available from very large to small capacities. Among them, the most commonly used pairs are lithium bromide-water (LiBr-H$_2$O) for a higher coefficient of performance (*COP*) and aqua ammonia (H$_2$O-NH$_3$) for producing cold at a lower temperature level [7]. A H$_2$O-NH$_3$ ARS is more complicated as it requires a rectifier mechanism to separate water vapor from ammonia, whereas the main problem for LiBr-H$_2$O ARS is the crystallization, which is also present for other salt-based aqueous solutions [8]. In recent years, numerous analyses have been undertaken to find better alternative working pairs of ARS, mostly using H_2O and NH_3 as refrigerants [9–12]. According to these studies, LiCl-H_2O seems to be one of the satisfying options as an absorption cycle working fluid for its advantages regarding the triple state point, long-term stability, comparatively lower cost, and better cycle performance, compared to LiBr-H_2O [13].

Numerous experimental studies on the properties of LiBr-H_2O and LiCl-H_2O solutions have been conducted and empirical formulations with reasonable accuracy are developed through curve fitting of the experimental data [14–18], which are helpful for researchers carrying out theoretical analyses of ARS using these working fluids. However, most of the studies focus on the evaluation of the LiBr-H_2O working pair and few studies are available on the analysis of LiCl-H_2O ARS. Parham et al. [19] stated that an absorption chiller using LiCl-H_2O has a ≈5–6% higher exergetic efficiency when the condensation temperature is 40 °C and absorber temperature is 35 °C, but a LiBr-H_2O working pair has a ≈0.6–0.8% higher *COP* under the same working conditions and even a ≈1.5–2% greater *COP* under the optimum conditions. Patel et al. [20] theoretically investigated a 1 RT (United States refrigeration ton) single effect LiCl-H_2O ARS on the basis of the first and second law of thermodynamics, and found that the *COP* of a LiCl-H_2O system is ≈4–9% higher, while the exergetic efficiency is ≈3–6% higher than that of LiBr-H_2O. Bellos et al. [21] parametrically examined the performance of a solar absorption cooling system under various heat sources and three ambient temperature levels and proved that the usage of the LiCl-H_2O pair performs better than LiBr-H_2O in terms of energetic and exergetic analysis for all the examined cases. Gogoi et al. [22] obtained a total of 34 different combinations of operating temperatures of single effect LiCl-H_2O ARS by using an inverse technique and specific optimization method. They found that LiCl-H_2O ARS is better than LiBr-H_2O ARS under the same conditions and the performance variation with the generator temperature in both systems solely depends on the operating temperatures of the condenser and absorber.

Almost all the cycles mentioned above contain just a single working pair and can only make use of partial energy in heat sources. For the effective utilization of the low-temperature thermal energy, She et al. [23] proposed a low grade heat-driven ARS combined with LiCl-H_2O and LiBr-H_2O working pairs in different pressure levels. Results show that the two parallel modes have much higher *COP* than the conventional double-stage LiBr-H_2O ARS in the specific operating ranges and the maximum *COP* improves by about 26.7% and 35%, respectively.

What is common with all the above studies is that the LiCl-H_2O is more desirable than LiBr-H_2O at a lower generator temperature and this is also confirmed in this paper. Besides, studies about LiCl-H_2O ARS are far from enough and many situations have not been examined yet. Meanwhile, in the parametric study of the absorption system, most studies simply focus on the influence of temperature on the main components of the thermodynamics performance, but rarely consider the actual heat transfer process between the pure water and the LiBr-H_2O or LiCl-H_2O solution. According to the studies, the temperature of each state point is given by assuming a constant increment or decrement with parameters of internal or external circuit flows in the whole process, such as a fixed temperature difference between the heat resource and generator, cooling water and condenser or absorber, chilled water and evaporator, as well as the hot and cold side, in the solution heat exchanger. Though it is a rational and convenient way, they do not accord with the actual situations very well. In the practical application, the operation of an absorption chiller will change with the environment conditions, resulting in deviations from the design parameters. By considering the internal heat transfer process in each component, the off-design performance of the absorption chiller could be predicted. In contrast to the above studies, parameters of all the state points will change as a whole in the absorption system to exhibit the influence of the outside environment on system operation.

For more precise models, a number of works have been carried out to explore the dynamic performance of absorption chillers. Ochoa [24,25] developed a mathematical model to conduct the dynamic analysis of a single-effect LiBr-H_2O absorption chiller by considering the correlations of the convective coefficients of absorption refrigeration processes and results showed the relative errors between experimental and numerical values were approximately 5% and 0.3% in the chilled and cold water circuits, respectively, when the variable overall coefficients were considered. Similarly, Olivier [26] presented the dynamic modeling of a 30 kW single effect LiBr-H$_2$O absorption chiller considering both the transient and steady state phases. The simulation results agreed very well with the experimental measurements with the mean absolute errors lower than 1 °C for the outlet temperatures of external water circuits in the components. In addition, Kohlenbach and Ziegler [27,28] built a dynamic model of absorption chiller based on internal and external steady-state enthalpy balances and took into account the mass and thermal storage terms as well as the solution transport delay in the components of the chiller. Results indicate that the deviations between the simulation and experimental data were approximately 1–3% in magnitude for the dynamic state and 0.7–3.5 K in temperature for steady state.

The above dynamic models with different approaches are applied to predict the reaction of the absorption chiller to the change of external conditions. However, the above studies created whole dynamic models that require specific information on the physical configurations of vessels for the heat exchanger surface area and characteristics of the inner and outer fluids in each component to calculate the heat transfer coefficient, which are hardly obtained from a commercial absorption chiller or supplied by manufacturers. In fact, the basic heat transfer characteristics in the components could be obtained from the design parameters. Different with the dynamic model, the observation of chiller's behaviors varying with time is not the aim of our work. The intention of the present study is to conduct a quick prediction of the off-design performance of the chiller under a steady state and examine the effects of external fluids on the system's operation. This is much easier to achieve by the new method with no need of detailed information or a complex mathematical model. More importantly, the dynamic models developed above are used only for conventional LiBr-H$_2$O absorption chillers, not for other alternative working pairs like LiCl-H$_2$O.

In another field of investigations, different methodologies have been proposed to analyze such commercial equipment by adding all operating characteristics of the chiller into a set of simple algebraic equations, which avoids the extensive information of the machine and complex numerical simulations. Hellmann et al. [29] developed a method using a so-called characteristic equation, which is able to approximate the part load behavior of single effect absorption chillers and heat pumps. In the following research, Puig [30] extended the adaptation of the characteristic equation method to double-effect commercial chillers. Gutiérrez-Urueta [31] obtained an extension of the characteristic equation method for adiabatic absorption based on a characteristic temperature difference, taking into account the facility features. However, these models use a multiple linear regression algorithm, where the accuracy relies on experimental or the manufacturer's data. Though the method requires less information about the chiller, it is limited to predictive the heat flow variation with temperature in a specific machine. The operating parameters of state points in the internal and external fluids are unable to be observed, which is available in our model. Furthermore, it is necessary to select the recorded data from the manufacturer's catalogue or measurements first, which have a great influence on the results. Thus, the characteristic equation method is merely appropriate for the existing commercial equipment and scarcely applied to LiCl-H$_2$O or other alternative working pairs, the same as the above dynamic model. Also, it hardly provides advice on the parameter or structure optimization of the chiller in the design. The commonality between this method and ours is that the simulations are based on steady-state conditions in the process.

This study developed a novel calculation method of ARS rarely used in other literatures. The model is based on energy and mass balances and the Dühring equation to describe the primary absorption cycle, and it is reliable and validated in the study of the LiCl-H$_2$O and LiBr-H$_2$O ARS simulations. Furthermore, the performance of LiCl-H$_2$O ARS has been examined comprehensively to complement the previous studies. Furthermore, this study is worthwhile because the variations of LiCl-H$_2$O absorption chiller with different operating conditions of external circuit flows (hot, cold, and chilled water) under design features have not been examined in detail before. A new computational procedure considering heat transfer characteristics of components has been created for comparative analysis of the whole LiCl-H$_2$O and LiBr-H$_2$O absorption chiller, something that is not well established for general application.

The computation work in the absorption cycle will be greatly reduced by combining a new method with a computational procedure because it avoids complex iterative calculations of solution properties. In addition, the developed model in this paper can be extended to more absorption cycles using other working pairs to predict their performances with high accuracy and efficiency.

2. System Description

Figure 1a displays the schematic diagram of a single-effect ARS and the process of the cooling effect generation driven by heat sources. For mathematical modeling of the system, its main components and state points are described using the pressure–temperature coordinates in Figure 1b. It involves two pressure levels: the generator (G) and condenser (C) operate at high pressure, and the absorber (A) and evaporator (E) work at low pressure level. The high temperature heat source (stream 11–12) adds heat in the generator to separate water vapor from the weak solution (stream 3). Then, the refrigerant vapor (stream 7) from the generator is condensed by the cooling water (stream 15–16) in the condenser. Afterwards, the liquid refrigerant (stream 8) flows through an expansion valve (EV2) to the evaporator and vaporizes at the saturation temperature to absorb heat from the chilled water (stream 17–18). Subsequently, the refrigerant vapor (stream 10) is conducted to the absorber, where it is absorbed by the strong solution (stream 6) going through the solution expansion valve (EV1) and the whole process is cooled down by cooling water (stream 13–14). Finally, the low concentration solution at the outlet of the absorber (stream 1) is pumped to the generator via a solution heat exchanger (SHE). The pressurized solution (stream 2) is preheated in the SHE by the high-temperature strong solution (stream 4–5) coming from the generator.

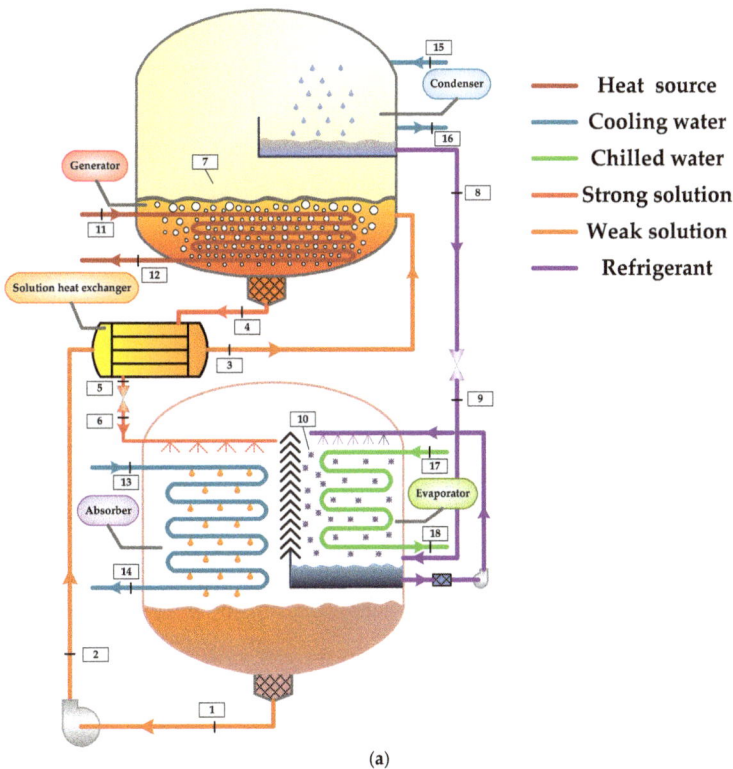

(a)

Figure 1. *Cont.*

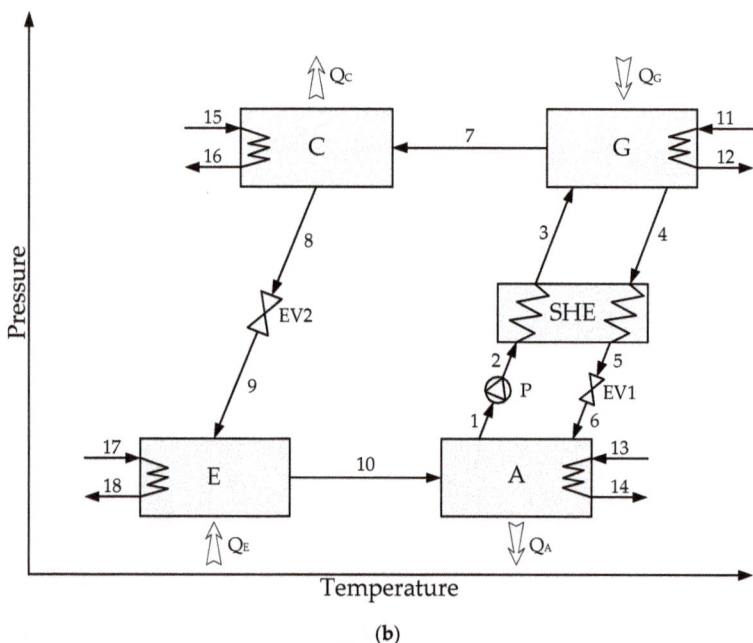

Figure 1. (**a**) Schematic representation of the single effect absorption refrigeration system, and (**b**) the pressure–temperature (P–T) diagram of the cycle.

3. Mathematical Modeling

3.1. Model Assumptions

In order to conduct the theoretical analysis of the model, some assumptions were made [12,19,22–24]. First, it was assumed that the entire system ran under steady state conditions and all the thermodynamic variables were homogenous within each component. Then, the solutions flowing out of the generator and absorber were considered to be saturated in equilibrium conditions at their respective temperature and pressure. Also, the refrigerant leaving the condenser and evaporator were in a saturated phase at the corresponding pressure. In addition, the pressure drop and heat loss in the components of the system were not considered, and neither was the work of the solution pump. At last, fluids passing through the expansion valves were regarded as adiabatic and isenthalpic. Additionally, the reference system was taken at 25 °C and 1 atmospheric pressure.

3.2. Thermodynamic Formulations

For the thermodynamic analysis of an ARS, the principles of mass and species conservation, as well as energy conservation including performance criteria, were applied. Each component of the chiller could be treated as a control volume with its inlet and outlet streams. The general equations of these principles must be fulfilled in all the components, where more specifics are presented as follows.

Mass balance:

$$\sum m_{i,in,k} - \sum m_{i,out,k} = 0 \tag{1}$$

Species balance:

$$\sum m_{i,in,k} x_{j,i,k} - \sum m_{i,out,k} x_{j,i,k} = 0 \tag{2}$$

Energy balance:

$$\sum H_{i,in,k} - \sum H_{i,out,k} + \sum Q_{u,k} - \sum W_k = 0 \tag{3}$$

$$\sum Q_{u,k} = \sum H_{u,in,k} - \sum H_{u,out,k} \tag{4}$$

Equations (1) and (2) represent the mass and species balances applied to the component k (G, A, C, E, SHE) and ensure that the amounts of each substance j (LiBr, LiCl, water) entering into component k equal the total amount of j leaving k. The energy balance in Equation (3) reflects that the difference between the inlet and outlet energy flows of stream i add the heat supplied by external utility u must be equal to the work done by the component k. In these equations, m indicates the mass flow rate of stream i (kg·s^{-1}), x denotes mass fraction of substance j in stream i (kg·kg^{-1}), H is the energy flow rate (kW), Q refers to the heat transfer capacity (kW), and W represents power (kW).

According to Equations (1) and (2), the mass balance equation and species balance equation in the generator and absorber can be expressed as the following Equations (5) and (6), respectively:

$$m_{ws} = m_{ss} + m_r \tag{5}$$

$$m_{ws}x_{ws} = m_{ss}x_{ss} \tag{6}$$

The circulation ratio CR is an important parameter, which is defined as the ratio of mass flow rate of weak solution coming from the absorber to the mass flow rate of the refrigerant:

$$CR = \frac{m_{ws}}{m_r} \tag{7}$$

According to Equations (5) and (6), Equation (7) can be expressed as:

$$CR = \frac{x_{ss}}{x_{ss} - x_{ws}} \tag{8}$$

In the thermodynamic analysis of the absorption cycle, we focus on the relationship between the concentration and temperature of solution in the generator or absorber and temperature of water in the condenser or evaporator under relevant pressure. In order to simplify the calculation process, the Dühring equation is recommended for dealing with the equilibrium temperature t of a solution and the dew temperature t_s at the corresponding pressure, shown as follows:

$$t^K = at_s^K + b \tag{9}$$

In the above equation, superscript K represents the Kelvin temperature scale, and a and b are the functions of solution composition. For LiBr-H$_2$O and LiCl-H$_2$O, a and b can be approximately expressed by a linear function of solution concentration in a moderately narrow concentration range. Then, Equation (9) can be rewritten as:

$$t^K = (a_0x + a_1)t_s^K + (b_0x + b_1) \tag{10}$$

where a_0, a_1, b_0, and b_1 are constants in the concentration range. The constants above are 0.538, 0.845, 48.3, and -35.6 in sequence for a LiBr-H$_2$O equilibrium solution according to Kim and Infante Ferreira's study [15]; likewise, 0.304, 0.988, 59.3, and -27.9 are used for LiCl-H$_2$O equilibrium solution based on Conde's study [14]. The absolute error is less than 1 K for LiBr-H$_2$O and 1.8 K for LiCl-H$_2$O within the operating range. Actually, the concentration of the solution varies in the range of $\approx$0.5–0.65 for LiBr-H$_2$O and $\approx$0.4–0.5 for LiCl-H$_2$O in the whole research process. Thus, the Dühring equation completely meets the requirement of computational accuracy.

In this paper, Kim and Infante Ferreira's method [32] is referred to when calculating the heat flows of components so as to facilitate the following iterative computations. This method is derived from Haltenberger's formula where the solution enthalpy at one concentration can be calculated from that of another concentration in the same solution temperature. Meanwhile, the partial specific enthalpy of the refrigerant takes an approximate value under a low-pressure condition. Finally, the heat load of each component can be expressed using the following formulas:

Generator:

$$Q_G = m_{ws}c_{p,ws}(t_4 - t_3) + m_r\left[\overline{X}_G h_{co} + c_{p,r}^v(t_7 - t_4)\right] \qquad (11)$$

where $\overline{X}_G$ is defined using:

$$\overline{X}_G = \left(\frac{t_{3a} + t_4}{2t_8}\right)^2 \frac{1}{a_0(x_{ws} + x_{ss})/2 + a_1} \qquad (12)$$

Absorber:

$$Q_A = m_{ws}c_{p,ws}(t_5 - t_1) + m_r\left[\overline{X}_A h_{ev} + c_{p,r}^v(t_{10} - t_1) - c_{p,ss}(t_5 - t_{6a})\right] \qquad (13)$$

where $\overline{X}_A$ is defined using:

$$\overline{X}_A = \left(\frac{t_{6a} + t_1}{2t_{10}}\right)^2 \frac{1}{a_0(x_{ws} + x_{ss})/2 + a_1} \qquad (14)$$

Condenser:

$$Q_C = m_r\left[h_{co} + c_{p,r}^v(t_7 - t_8)\right] \qquad (15)$$

Evaporator:

$$Q_E = m_r\left[h_{ev} - c_{p,r}^l(t_8 - t_{10})\right] \qquad (16)$$

Solution heat exchanger:

$$Q_{SHE} = (m_{ws} - m_r)c_{p,ss}(t_4 - t_5) = m_{ws}c_{p,ws}(t_3 - t_2) \qquad (17)$$

In order to make it more precise for application, some empirical formulations used to describe the thermodynamic properties of working pairs are adopted by this model instead of constant values. The latent heat of evaporation and condensation in different temperatures are calculated from Wagner et al. [33]. The specific heat of LiBr-H_2O and LiCl-H_2O solutions are computed using the correlations of Patek and Klomfar [17,18].

The system energetic performance is represented in terms of *COP*. It is defined as the ratio of cooling capacity to the heat input and is presented in Equation (18) [19,21]:

$$COP = \frac{Q_E}{Q_G} \qquad (18)$$

The Carnot coefficient of performance (*COP_c*) represents the maximum possible coefficient of the ARS, as shown in Equation (19) [34]:

$$COP_c = \left(\frac{t_4^K - t_1^K}{t_4^K}\right)\left(\frac{t_{10}^K}{t_8^K - t_{10}^K}\right) \qquad (19)$$

By comparing the *COP* with *COP_c*, the efficiency ratio is given as:

$$\eta_{eff} = \frac{COP}{COP_c} \qquad (20)$$

The exergetic efficiency of the chiller is the ratio of the exergy produced in the evaporator to the exergy supplied to the generator. It can be written as [21]:

$$\eta_{ex} = \frac{Q_E\left|\left(1 - \frac{t_0^K}{t_E^K}\right)\right|}{Q_G\left(1 - \frac{t_0^K}{t_G^K}\right)} \qquad (21)$$

3.3. Standard Design of Absorption Chiller with LiBr-H₂O/LiCl-H₂O Working Pair

This section intends to design the single effect absorption chiller using LiBr-H$_2$O and LiCl-H$_2$O. The main components of the absorption chiller (generator, absorber, condenser, evaporator, and solution heat exchanger) are virtually five heat exchangers. The objective of the design is to determine the heat transfer characteristics of each component and mass flow rate of each state point under the condition that the cooling capacity and design parameters in both chillers are the same.

The heat transfer model of the heat exchanger in each component (*k*) can be described by the overall heat transfer coefficient-area product (*UA*) and the logarithmic mean temperature difference (Δt^{lm}), shown as follows:

$$Q_k = (UA)_k \Delta t_k^{lm} \tag{22}$$

In the above formula, the logarithmic mean temperature difference can be calculated conveniently with high-precision by employing Chen's approximation [35,36], which is described as a function of temperature differences between the hot and cold end, as shown in the following equation:

$$\Delta t_k^{lm} \cong \left[\Delta t_k^h \Delta t_k^c \frac{\Delta t_k^h + \Delta t_k^c}{2} \right]^{\frac{1}{3}} \tag{23}$$

According to the energy balance equations shown in Equations (3) and (4), heat transfer rates of the main components can be expressed in terms of energy flow rates of external utilities, as in the following Equations (24)–(27):

$$Q_G = m_{11} c_{p,w} (t_{11} - t_{12}) \tag{24}$$

$$Q_A = m_{13} c_{p,w} (t_{14} - t_{13}) \tag{25}$$

$$Q_C = m_{15} c_{p,w} (t_{16} - t_{15}) \tag{26}$$

$$Q_E = m_{17} c_{p,w} (t_{17} - t_{18}) \tag{27}$$

In this study, the rated cooling load of the absorption chiller was 50 kW. The effectiveness of heat exchanger was set at 0.7. To carry out the design of the absorption chiller, the temperature of each state point in the operating conditions should be determined at first. An absorption chiller contains three external circuit flows: hot water, cooling water, and chiller water. The states of these fluids are closely related to the actual environment, especially inlet temperatures. Once the inlet temperatures of external fluids are selected, temperatures of the other points can be typically determined by some increments or decrements. In accordance with References [22,23,34], the design specifications applied in this paper for both systems are shown in Table 1:

Table 1. Design temperatures of state points in the components.

Component	State Point	Symbol	Value
Generator	Inlet temperature of hot water (°C)	t_{11}	90
	Outlet temperature of hot water (°C)	$t_{12} = t_{11} - 7$	83
	Outlet solution temperature from generator (°C)	$t_4 = t_{12} - 3$	80
Absorber	Inlet temperature of cooling water (°C)	t_{13}	30
	Outlet temperature of cooling water (°C)	$t_{14} = t_{13} + 5$	35
	Outlet solution temperature from absorber (°C)	$t_1 = t_{14} + 3$	38
Condenser	Inlet temperature of cooling water (°C)	t_{15}	30
	Outlet temperature of cooling water (°C)	$t_{16} = t_{15} + 5$	35
	Condensation temperature (°C)	$t_8 = t_{16} + 3$	38
Evaporator	Inlet temperature of chilled water (°C)	t_{17}	13
	Outlet temperature of chilled water (°C)	$t_{18} = t_{17} - 5$	8
	Evaporation temperature (°C)	$t_{10} = t_{17} - 7$	6

After setting up these temperatures, the mass flow rate of each point, as well as the heat transfer characteristics and heat load of each component, could be determined from the above equations. In most studies, temperature differences between the fluids in Table 1 are treated as constant to research the operation performance of an absorption chiller, which is in fact at variance with the actual situation. In the next section, a new effective method has been developed to overcome this problem.

3.4. Calculation Procedure of Absorption Chiller under Design Features

In the control volume of the absorption chiller, heat transfers of components take place between the external circuit flows (hot, cold, and chilled water) and its internal circuit flows (the refrigerant and the LiBr-H_2O/LiCl-H_2O solution). It is more practical to evaluate the influences of external fluids on the absorption chiller. In order to build a model that could be commonly used, the heat transfer characteristics (UA) of all the components are assumed to be invariable within the entire operating range, just as in References [12,37–39]. According to the research, selecting the design parameters is adequate to simulate the real situation with a good approximation because the overall heat transfer coefficients vary within a narrow range most of the time [12]. In particular for this study, the mass flow rates of the solution pump and external circuit flows are kept constant in an attempt to minimize its effects on the overall heat transfer coefficients, so it allows for comparison of the results obtained from the LiBr-H_2O and LiCl-H_2O absorption chillers. A computer program in Matlab language (Matlab R2014a, The MathWorks Inc., Natick, MA, America) has been made to carry out the simulation. Figure 2 shows a flow chart of the computer program (see in Supplementary Materials).

As shown in the simulation procedure, it contains a five-step iteration of different temperatures corresponding to the main components of the chiller. In this way, it is effective to describe the operation of chiller and evaluate the influences of external circuits on its overall performance. According to the studies on thermodynamic properties of the working pairs, the vapor pressure of solution (P) is calculated using an empirical formulation, which is described as the function of solution temperature (t) and mass fraction (x), i.e., $P = f(t, x)$. In the mathematical modeling of the system, the vapor pressure of the solution equals to the saturation pressure of water in the condenser or evaporator, $P = f(t_s)$. As a result, in the most cases, the temperature and pressure of solution are acquired to solve its mass fraction, $x = f(P, t)$. The complex multivariable nonlinear equations hinder the iterative computation of the solution parameter immensely.

The method in this study using Dühring equation shown as Equation (10) completely avoids the problem and makes it much easier to deal with the relation between the solution temperature and mass fraction. Besides, in the heat transfer model, the logarithmic mean temperature difference is calculated using Equation (23) with a high accuracy. In this process, the logarithmic equation is converted into an algebraic equation, which provides a simple way to solve for the fluid temperature. Consequently, the developed method in this study will reduce the amount of computation compared with the conventional enthalpy-based calculation. In conclusion, the established model is efficient and convenient to predict the performance of an absorption chiller, especially for alternative working pairs without specific information.

Start

Input given data: m_{11}, m_{13}, m_{15}, m_{17}, m_1, UA_G, UA_A, UA_C, UA_E, UA_S

Input data: t_{11}, t_{13}, t_{15}, t_{17}

Assume value of t_{10} ◄— D

Calculate the latent heat of evaporation $h_{ev}=f(t_{10})$

Assume value of t_8 ◄— C

Calculate the latent heat of condensation $h_{co}=f(t_8)$

Assume value of t_4 ◄— B

Solve x_{ss} and calculate t_{6a} according to Eq.(10)

Assume value of t_5 ◄— A

Calculate $c_{p,5}$

Assume value of t_1 ◄—

Solve x_{ws} according to Eq.(10). Calculate the CR, m_r, t_{3a}, $\bar{\chi}_G$, $\bar{\chi}_A$, Q_A, solve t_{14} from Eq.(25)

Determine new t_1' according to Eq.(22), $\varepsilon=|t_1'-t_1|$

$\varepsilon<10^{-2}$ — No

Yes

Calculate Q_{SHE} and solve t_3 from Eq.(17)

Determine new t_5' according to Eq.(22), $\varepsilon=|t_5'-t_5|$

$\varepsilon<10^{-2}$ — No → A

Yes

Calculate Q_G from Eq.(11) and solve t_{12} from Eq.(24)

Determine new t_4' according to Eq.(22), $\varepsilon=|t_4'-t_4|$

$\varepsilon<10^{-2}$ — No → B

Yes

Calculate Q_C from Eq.(15) and solve t_{16} from Eq.(26)

Determine new t_8' according to Eq.(22), $\varepsilon=|t_8'-t_8|$

$\varepsilon<10^{-2}$ — No → C

Yes

Calculate Q_E from Eq.(16) and solve t_{18} from Eq.(27)

Determine new t_{10}' according to Eq.(22), $\varepsilon=|t_{10}'-t_{10}|$

$\varepsilon<10^{-2}$ — No → D

Yes

Print all the output results

End

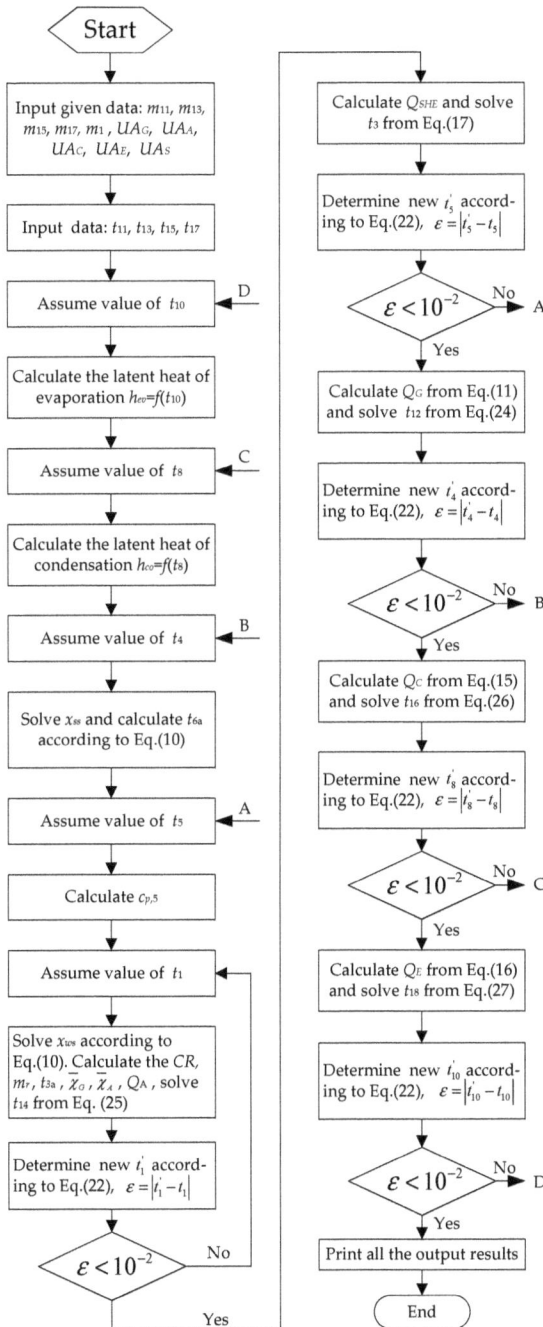

Figure 2. Flow chart of the calculation procedure.

4. Results and Discussion

4.1. Validation of the Model

In order to validate the developed model, the simulation results are compared with related research reported by Anand and Kumar [40] and Kaushik and Arora [41] for a LiBr-H$_2$O system, and Patel and Pandya [20] for a LiCl-H$_2$O system. In the comparison, the values of input parameters (mainly the temperatures of the components and mass flow rate of the refrigerant) in the examined cases were identical. According to the references, for LiBr-H$_2$O system, the input values were set as: $t_G = 87.8\,°C$, $t_A = t_C = 37.8\,°C$, $t_E = 7.2\,°C$, $\varepsilon = 0.7$, $m_r = 1\,kg·s^{-1}$. In the LiCl-H$_2$O system, the parameters were as follows: $t_G = 78\,°C$, $t_A = 40\,°C$, $t_C = 36\,°C$, $t_E = 10\,°C$, $\varepsilon = 0.7$, $m_r = 0.00148\,kg·s^{-1}$.

It was observed that the simulation results of the present work agreed well with the references' data and the differences between them were within ±1% for a LiBr-H$_2$O system and ±3% for a LiCl-H$_2$O system. In the two systems, the maximum difference percentage of heat capacity all occurred in the absorber (0.8% and 2.6% respectively), which produces no effect on *COP*. The details are presented in Table 2. It proves that the calculation model is valid and reliable.

Table 2. Comparisons of energy analysis of present work with references' data.

Name	Symbol	LiBr-H$_2$O					LiCl-H$_2$O		
		Anand [40]	Kaushik [41]	Present Work	Difference Percentage (%) with Anand	Difference Percentage (%) with Kaushik	Patel [20]	Present Work	Difference Percentage (%) with Patel
Heat capacity of generator (kW)	Q_G	3073.11	3095.70	3074.45	0.04	−0.69	4.465	4.519	1.21
Heat capacity of absorber (kW)	Q_A	2922.39	2945.27	2945.90	0.80	0.02	4.268	4.379	2.6
Heat capacity of condenser (kW)	Q_C	2507.89	2505.91	2504.22	−0.15	−0.07	3.705	3.691	−0.38
Heat capacity of evaporator (kW)	Q_E	2357.17	2355.45	2355.93	−0.05	0.02	3.517	3.505	−0.34
Coefficient of performance	*COP*	0.7670	0.7609	0.7663	−0.09	0.71	0.7877	0.7758	−1.51

4.2. Sensitivity Analysis of the Design Parameters

From the results and mathematical models in previous studies, it indicates that the performance of a single effect absorption system mainly depends on the selection of main design parameters such as absorber temperature t_1, generator temperature t_4, condenser temperature t_8, evaporator temperature t_{10}, and heat exchanger effectiveness [34].

In this section, the sensitivity analysis of these parameters is conducted to examine the effect on LiBr-H$_2$O and LiCl-H$_2$O ARS performances. The method of control variables was employed for the sensitivity analysis, where one of the system parameters is varied while the other parameters were kept constant. The purpose of this simulation was to predict the most efficient pair and determine the operating temperature range. The results are shown in the following (see in Supplementary Materials).

4.2.1. Effect of Generator Temperature on System

Figures 3 and 4 present the variations of the coefficient of performance (*COP*) and efficiency ratio (η_{eff}) with generator temperature (t_4) in LiBr-H$_2$O and LiCl-H$_2$O systems. The system performance was analyzed under three different condenser and absorber temperature levels (32 °C, 36 °C, and 40 °C) in order to cover a greater range of operating conditions. In this process, the absorption temperature was set equal to condensation temperature ($t_1 = t_8$).

As shown in Figures 3 and 4, it is obvious that the LiCl-H$_2$O working pair displayed better performances than LiBr-H$_2$O in the operating range of examined cases. It can be observed that both the *COP* and efficiency ratio value increased sharply with generator temperature initially, but *COP* tended to level off and even drops somewhat rather than show an apparent decline in the efficiency ratio curves with a further increase in generator temperature. Moreover, a lower condensation and absorption temperature led to a greater *COP* in the same generator temperature and the maximum value approximated to 0.8 for both working pairs. These curves reflect that there existed a minimum generator temperature in every case that brought about an adequate *COP* in the application of absorption system. Furthermore, it can be inferred that this minimum generator temperature was closely related to the maximum efficiency ratio of system, as the calculated results show.

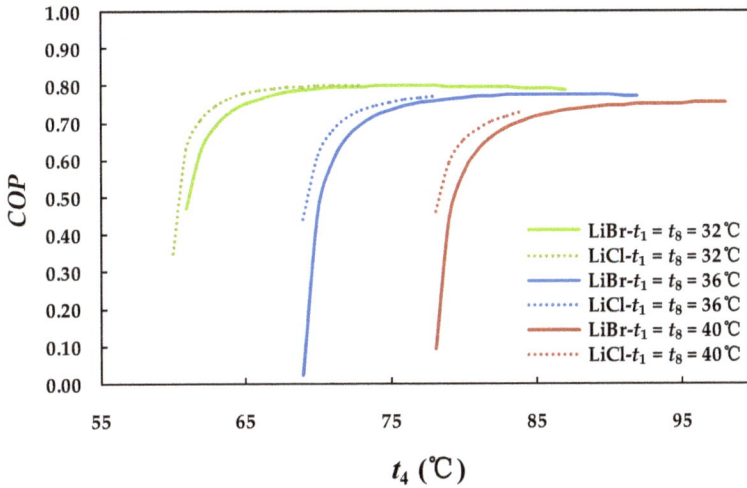

Figure 3. Generator temperature impact on the *COP* of the absorption system under various condenser and absorber temperature levels (32 °C, 36 °C, and 40 °C) for the two working pairs ($t_{10} = 7$ °C, $\varepsilon = 0.7$).

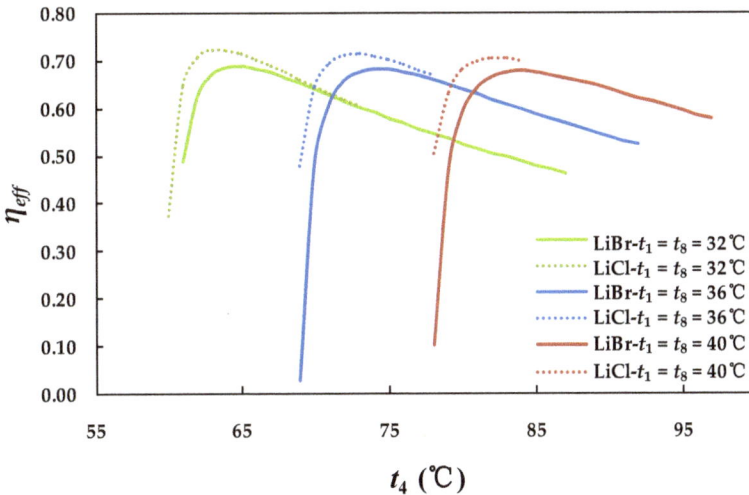

Figure 4. Generator temperature impact on the efficiency ratio of absorption system under various condenser and absorber temperature levels (32 °C, 36 °C, and 40 °C) for the two working pairs ($t_{10} = 7$ °C, $\varepsilon = 0.7$).

In Figure 4, each curve of the efficiency ratio in the two systems had almost the same tendency, no matter the limitations, just like the curves moving in the direction of generator temperature increasing, and there existed almost the same maximum values with 0.68 for LiBr-H$_2$O and 0.7 for LiCl-H$_2$O for an intermediate generator temperature for each curve. The reason for this maximum efficiency ratio is the system COP_c increased with the generator temperature, but the COP curve almost flattened.

It is also useful to state that higher generator temperatures must match higher condensation and absorption temperatures because of the solubility limitation of the two working pairs. Although it has a greater performance than the LiBr-H$_2$O system, the LiCl-H$_2$O system is only allowed to operate in a much smaller range of working conditions, which primarily limits its application.

4.2.2. Effect of Absorber and Condenser Temperatures on System

The effects of the variation in the absorber and condenser temperatures on the system COP and efficiency ratio are illustrated in Figures 5 and 6. It can be observed that LiCl-H$_2$O was the working pair with the higher performance but had an extreme operating range that was suitable only for higher condensation and absorption temperatures, and the performance declined significantly with the temperature increasing. Conversely, LiBr-H$_2$O system was more adaptable for working conditions varying in a large range. When the generator temperature was kept at 80 °C, the LiCl-H$_2$O absorption cycle could only run in the condenser and absorber temperatures range of 38–41 °C, compared with 27–40 °C in the LiBr-H$_2$O absorption cycle. Especially, the COP of the LiBr-H$_2$O system kept basically in line at a high level close to 0.8 for all cases under a lower condensation and absorption temperature no matter the generator temperature changes. However, the curves of the efficiency ratio were clearly different, as shown in Figure 6, because they were maximized for a specific generator temperature in every case. In the examined cases, the maximum values of the efficiency ratio were nearly the same, which was about 0.67 for the LiBr-H$_2$O system and 0.70 for the LiCl-H$_2$O system.

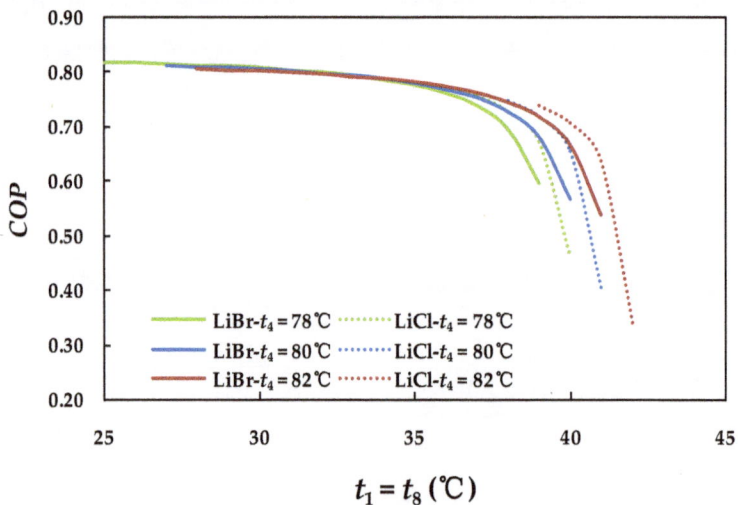

Figure 5. Condenser and absorber temperatures impact on the COP of the absorption system under various generator temperature levels (78 °C, 80 °C, and 82 °C) for the two working pairs (t_{10} = 7 °C, ε = 0.7).

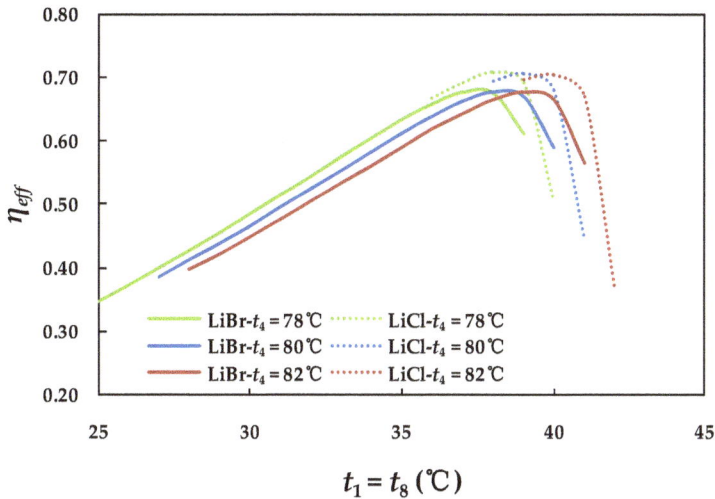

Figure 6. Condenser and absorber temperatures impact on the efficiency ratio of the absorption system under various generator temperature levels (78 °C, 80 °C, and 82 °C) for the two working pairs ($t_{10} = 7$ °C, $\varepsilon = 0.7$).

4.2.3. Effect of Absorber and Condenser Temperatures on System

Figures 7 and 8 depict the effect of the condenser or absorber temperature on *COP* and the efficiency ratio in both systems. It can be seen that a decrease in *COP* and the efficiency ratio occurred when increasing either of the temperatures. Moreover, the performances of both systems degraded more excessively in the high temperature zone.

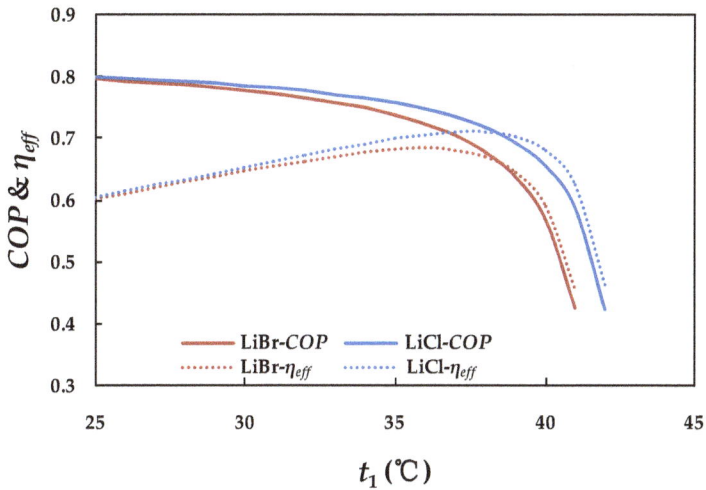

Figure 7. Absorber temperature impact on the *COP* and efficiency ratio of the absorption system under the examined operating condition for the two working pairs ($t_4 = 80$ °C, $t_8 = 40$ °C, $t_{10} = 7$ °C, $\varepsilon = 0.7$).

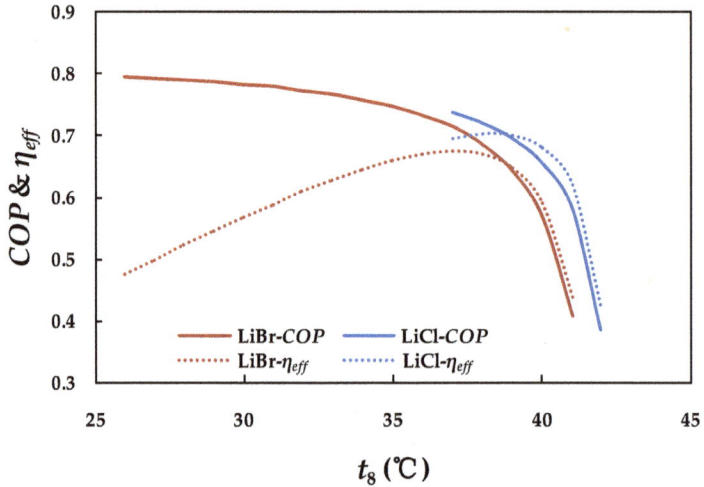

Figure 8. Condenser temperature impact on the *COP* and efficiency ratio of absorption system under the examined operating condition for the two working pairs ($t_1 = 40\,°C$, $t_4 = 80\,°C$, $t_{10} = 7\,°C$, $\varepsilon = 0.7$).

Comparing the results of two working pairs from Figures 7 and 8, it is shown that, for the LiCl-H$_2$O pair, the absorber temperature was allowed a wide range when condenser temperature was set at 40 °C, which was not true for the absorber temperature. Therefore, a conclusion can be reached that the condenser temperature had the greater impact on the LiCl-H$_2$O system. If the condenser temperature was too small, the system had operating trouble because the concentration of strong solution got larger, which led to solution crystallization. On the other hand, the concentration of strong solution became lower with the condenser temperature increasing, which brought about the solution cycle ratio increase at a constant cooling capacity, and eventually resulted in the degradation of system performance for the heat load of a generator rising rapidly. There also existed a maximum value in each of the efficiency ratio curves and the corresponding temperatures in LiCl-H$_2$O system were higher. This suggests that the LiCl-H$_2$O working pair was superior to LiBr-H$_2$O under the condition with a relatively higher condenser or absorber temperature. However, both of the systems were sensitive to overly high temperatures in the condenser or absorber, and the performance declined sharply. As such, the cooling water temperature should be strictly controlled to maintain a proper condenser and absorber temperature.

4.2.4. Effect of Evaporator Temperature on System

The variations of *COP* with evaporator temperature in the two systems are shown in Figure 9. It can be seen that the *COP* of both systems increased with evaporator temperature, but the growth rate was weakened and all the values approached 0.8 eventually. At this point, it is also important to state that a higher generator temperature was demanded to generate a lower evaporator temperature effect because the cooling production was harder to be achieved with low grade energy. Moreover, the *COP* was greater in the higher generator temperature at the same evaporator temperature level.

In Figure 10, it can be seen that each curve of the efficiency ratio had a maximum value according to the generator temperature level. To explain the maximum points of the efficiency ratio curves, a higher evaporating temperature or generator temperature level led to a higher COP_c according to Equation (19) and the growth rate of COP_c was greater than *COP* with the evaporating temperature increasing for each curve. At the higher evaporator temperature condition, the COP_c played the dominant role in the efficiency ratio because the *COP* values were almost the same in these cases.

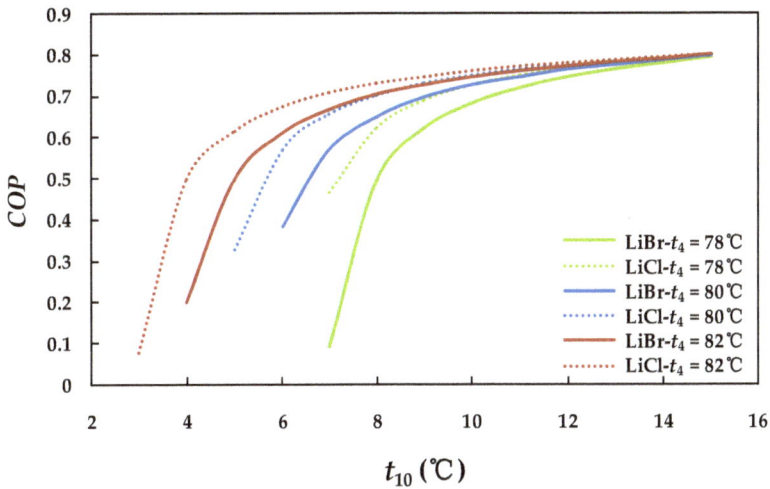

Figure 9. Evaporator temperature impact on the *COP* of the absorption system under various generator temperature levels (78 °C, 80 °C, and 82 °C) for the two working pairs ($t_1 = t_8 = 40$ °C, $\varepsilon = 0.7$).

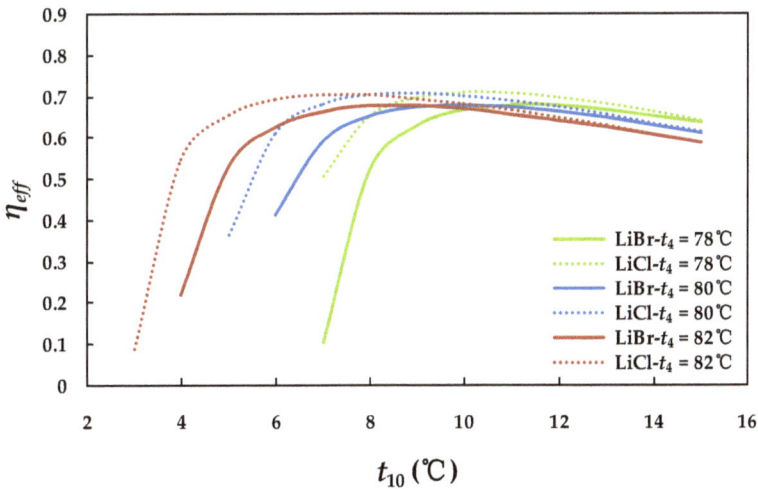

Figure 10. Evaporator temperature impact on the efficiency ratio of absorption system under various generator temperature levels (78 °C, 80 °C, and 82 °C) for the two working pairs ($t_1 = t_8 = 40$ °C, $\varepsilon = 0.7$).

4.2.5. Effect of Effectiveness of Solution Heat Exchanger on System

Figures 11 and 12 present the system performance in terms of *COP* and efficiency ratio varying with the solution heat exchanger effectiveness in the two systems. It can be seen that the effectiveness of the solution heat exchanger had a great impact on the system performance, especially at the lower generator temperature. This makes sense because the higher solution heat exchanger effectiveness improved the heat transfer between the generator and absorber and decreased the outside heat input and dissipation. Besides, LiCl-H_2O system had a great advantage over LiBr-H_2O at a lower generator temperature level, and the results agree with previous conclusions.

The performances of the two absorption cycles go up steadily with the heat exchanger effectiveness increasing, but to a markedly larger extent in the higher section. The tendency of the efficiency ratio is similar to the *COP* curve because the COP_c value was nearly at 1 in all examined cases. With the increase of the generator temperature, the gap between the two systems diminished. It is also obvious that the temperature of the weak solution leaving the solution heat exchanger decreased as the effectiveness rose, which indicates that the chances of solution crystallization increased.

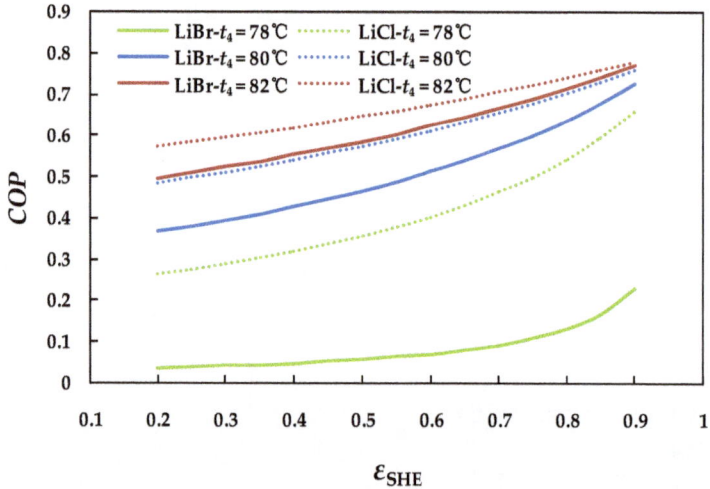

Figure 11. Heat exchanger effectiveness impact on the *COP* of the absorption system under various generator temperature levels (78 °C, 80 °C, and 82 °C) for the two working pairs ($t_1 = t_8 = 40$ °C, $t_{10} = 7$ °C, $\varepsilon = 0.7$).

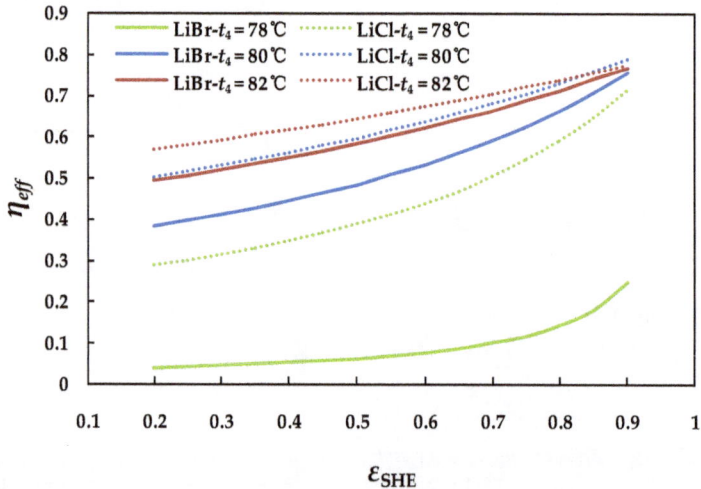

Figure 12. Heat exchanger effectiveness impact on the efficiency ratio of the absorption system under various generator temperature levels (78 °C, 80 °C, and 82 °C) for the two working pairs ($t_1 = t_8 = 40$ °C, $t_{10} = 7$ °C, $\varepsilon = 0.7$).

4.3. Analysis of the Two Absorption Chillers at Off-design Conditions

In the actual operation of a designed absorption chiller, the working conditions of the external circuit flows varies with the ambient conditions and the performance and each state point parameter of the chiller will change. Thus, it is necessary to examine the influence of the external circuit flows on the chiller and observe its regularity for practical application. In this work, the approach for the cycle simulation was carried out to establish a set of state points of the system and change the relevant parameters of external circuit fluids around it. Furthermore, the design features, particularly the heat transfer characteristics of the heat exchangers, had a decisive effect on the absorption chiller performance. The operating parameters of all the state points used for the initial values of simulation are listed in Table 3.

Table 3. Single effect absorption chiller operating parameters.

Point	LiBr-H_2O			LiCl-H_2O		
	t (°C)	m (kg·s^{-1})	x (%)	t (°C)	m (kg·s^{-1})	x (%)
1	38.0	0.50039	55.858	38.0	0.25584	43.874
2	38.0	0.50039	55.858	38.0	0.25584	43.874
3	65.4	0.50039	55.858	64.6	0.25584	43.874
4	80.0	0.47914	58.335	80.0	0.23459	47.849
5	50.6	0.47914	58.335	50.6	0.23459	47.849
6	50.6	0.47914	58.335	50.6	0.23459	47.849
7	80.0	0.02125	0	80.0	0.02125	0
8	38.0	0.02125	0	38.0	0.02125	0
9	6.0	0.02125	0	6.0	0.02125	0
10	6.0	0.02125	0	6.0	0.02125	0
11	90.0	2.43054	0	90.0	2.34078	0
12	83.0	2.43054	0	83.0	2.34078	0
13	30.0	3.29257	0	30.0	3.17298	0
14	35.0	3.29257	0	35.0	3.17298	0
15	30.0	2.53054	0	30.0	2.53054	0
16	35.0	2.53054	0	35.0	2.53054	0
17	13.0	2.39232	0	13.0	2.39232	0
18	8.0	2.39232	0	8.0	2.39232	0

From the above operating parameters of each state point, standard working conditions of the two absorption chillers can be obtained from thermodynamic analysis, such as heat capacity (Q) and heat transfer characteristics (UA) of each component, as presented in Table 4. It can be seen that the heat load in the condenser was slightly higher than that in the evaporator and that was also higher for the generator and absorber. This was primarily due to the superheating vapor in the generator and condenser compared with the saturated vapor in the absorber and evaporator. The highest heat transfer rate occurred in the generator that was approximately 71.12 kW for LiBr-H_2O and 68.49 kW for LiCl-H_2O. As a result, the LiCl-H_2O system had a higher *COP* than the LiBr-H_2O system for the same design parameters.

Table 4. Single effect absorption chiller performance parameters and design data.

Items	LiBr-H_2O	LiCl-H_2O
Heat transfer rate of component (kW)		
Generator (Q_G)	71.12	68.49
Absorber (Q_A)	68.81	66.32
Condenser (Q_C)	52.89	52.89
Evaporator (Q_E)	50	50
Solution heat exchanger (Q_{SHE})	27.29	18.45
COP	0.703	0.730

Table 4. *Cont.*

Items	LiBr-H$_2$O	LiCl-H$_2$O
Heat transfer characteristics (kW·°C^{-1})		
Generator (UA_G)	5.287	4.973
Absorber (UA_A)	6.049	5.829
Condenser (UA_C)	10.387	10.387
Evaporator (UA_E)	12.566	12.566
Solution heat exchanger (UA_S)	2.009	1.323

4.3.1. Effect of Hot Water Inlet Temperature in Generator

Figure 13 shows the variation of *COP* and exergetic efficiency with hot water inlet temperature. The *COP* curves of both chillers increased in the first stage but degraded a little after an intermediate temperature of about 98 °C for the LiBr-H$_2$O system and 83 °C for the LiCl-H$_2$O system, which were also the maximum values. However, the exergetic efficiency of the chiller was negatively affected by the increase of the heat source inlet temperature, dropping from about 0.25 to 0.18 for the LiBr-H$_2$O system and 0.27 to 0.21 for the LiCl-H$_2$O system in their own operating ranges.

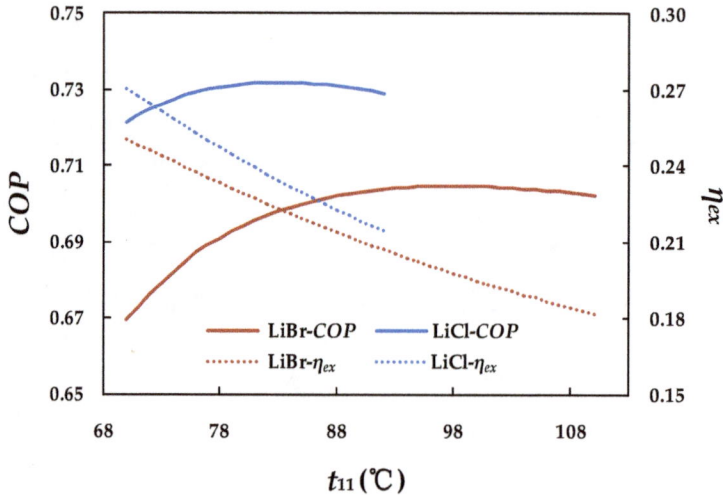

Figure 13. Variation of *COP* and exergy efficiency with the hot water inlet temperature for the two absorption chillers.

According to Figures 14 and 15, the inlet temperature of the heat source started from 70 °C and rose to 110 °C, but LiCl-H$_2$O system ends up at 92 °C due to a crystallization problem. With the increase in inlet temperature of the heat source water, the exit temperatures of the internal and external fluids in the main components showed close to linear growth except in the evaporator for both chillers. This was because temperatures of the refrigerant and the strong solution at the outlet of the generator increased as the heat source temperature went up, which led to the average temperatures of both condenser and the absorber increasing. Thus, for the two components, the temperature differences of heat transfer between the solution and water increased under the inlet cooling water being constant, which resulted in heat transfer rates of these components improving. As a consequence, the exit temperatures of the cooling water in the condenser and the absorber increased with constant mass flow rates and inlet temperatures. Besides, the increase of the refrigerant quantity augmented the cooling capacity in the evaporator, which accounted for the temperature of the outlet chilled water decrease because the mass flow rate and inlet temperature of chilled water were invariable. According to Equation (22), the evaporating temperature decreased eventually and also caused the corresponding low pressure level.

When the inlet temperature of the heat source water increased by 1 °C, the cooling capacity of the chiller increased by ≈1.2%–4.1% for the LiBr-H₂O system and ≈1.9%–3.9% for the LiCl-H₂O system, and the higher the temperature of the heat source, the slower the growth rate of the cooling capacity.

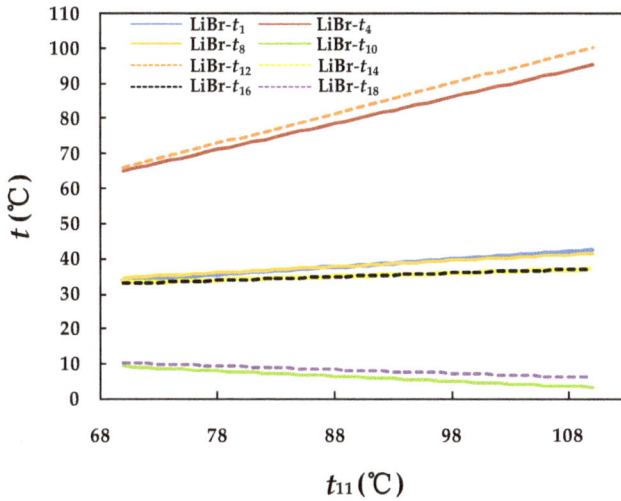

Figure 14. Variations of the exit temperatures of the internal and external fluids with hot water inlet temperature for the LiBr-H₂O absorption chiller.

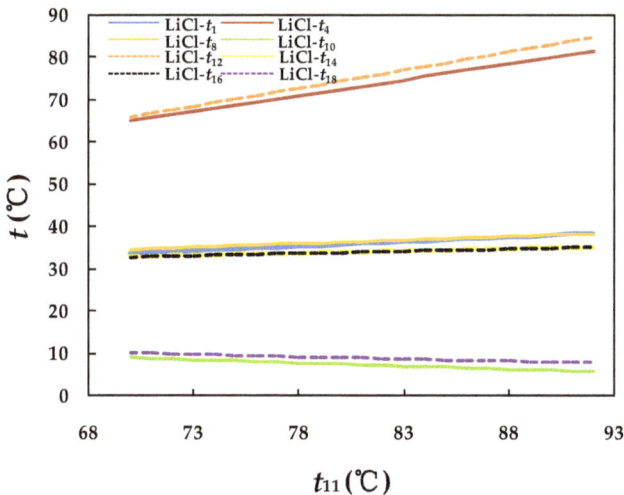

Figure 15. Variations of the exit temperatures of the internal and external fluids with hot water inlet temperature for the LiCl-H₂O absorption chiller.

4.3.2. Effect of Cooling Water Inlet Temperature in Absorber and Condenser

It can be ascertained from Figure 16 that cooling water between the condenser and absorber severely affected the performance of both absorption chillers. The LiCl-H₂O system could not operate at a lower cooling temperature than 26 °C. The *COP* and exergetic efficiency displayed an initial steady degression, but plummeted when the inlet temperature of the cooling water passed 41 °C. This was mainly because the cooling capacity slipped to a smaller value and descended quickly compared to the heat input at the generator as the inlet temperature rose. In the meantime, it can be observed from Figures 17 and 18 that the temperatures of components all displayed an upward trend while the heat duties of all components were completely the opposite. The reason will be explained as follows.

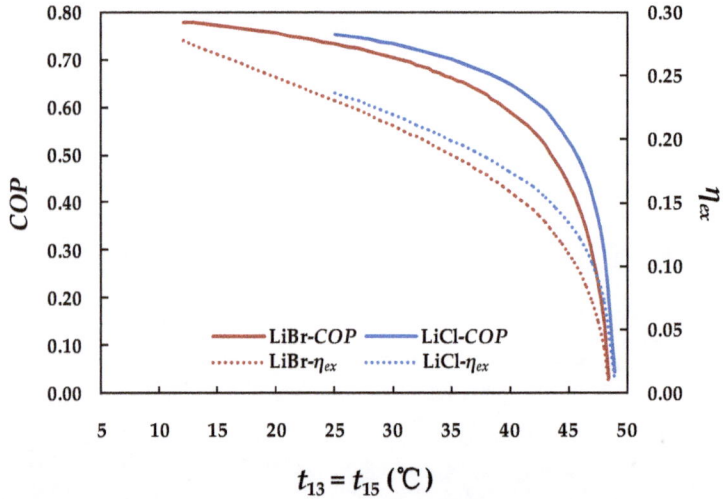

Figure 16. Variation of *COP* and the exergy efficiency with the cooling water inlet temperature for the two absorption chillers.

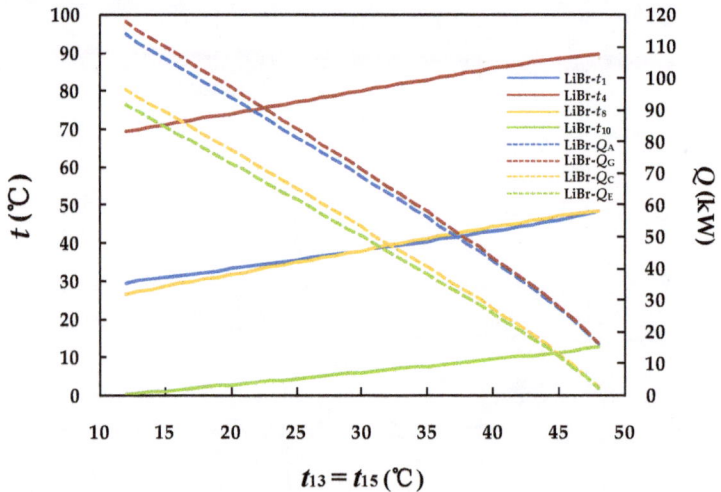

Figure 17. Variation of the exit temperature and heat capacity of each component with the cooling water inlet temperature for the LiBr-H$_2$O absorption chiller.

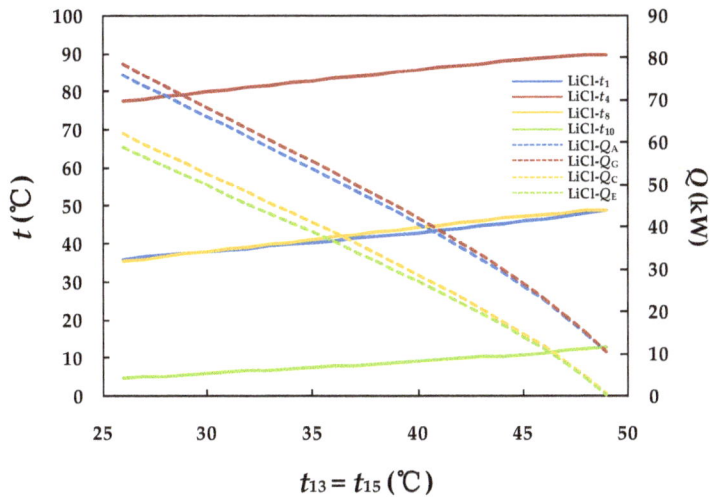

Figure 18. Variation of the exit temperature and heat capacity of each component with the cooling water inlet temperature for LiCl-H$_2$O absorption chiller.

In the first place, the working conditions of the absorber and condenser were directly affected by the inlet temperature of the cooling water. On the one hand, the weak solution temperature of the absorber increased with the inlet cooling water temperature increasing, which led to the increase of the mass fraction of the weak solution. On the other hand, the condensation temperature went up due to the enhanced heat transfer with the cooling water, and also the corresponding condensation pressure, causing a reduction of the strong solution concentration. As a result, the circulation ratio of the solution increased while the mass flow rate of the solution pump remained constant, resulting in the reduction of the refrigerant quantity and eventually the condensing and cooling capacity decreased. Consequently, the heat duties of the generator and absorber declined due to a lower solution circulation ratio; therefore, the temperature difference between the inlet and outlet temperature of the external circuit water decreased under the condition that mass flow rates were unchanged.

The rate of decrease in the cooling capacity was within 10% for the LiBr-H$_2$O system with cooling water temperatures between 12 °C and 41 °C in comparison to the LiCl-H$_2$O system between 26 °C and 41 °C; otherwise, the rate dropped dramatically above this range and even more than 40% at the highest temperatures.

4.3.3. Effect of Chilled Water Inlet Temperature in Evaporator

The effects of the inlet temperature of chilled water in the evaporator on the performance of both absorption chillers are illustrated in Figures 19–21. It can be seen that the *COP* and heat capacities of all the components present a linear growth, as well as the evaporator temperature, which is exactly contrary to the tendency of the exergetic efficiency and the solution temperature at the exit of the generator. Likewise, the temperatures of the absorber and condenser were much the same for both systems and increased slightly with the inlet temperature of chilled water, which were a little higher for the absorber temperature.

In order to explain them, the effect of the chilled water on the evaporator should be investigated first. It was not difficult to find that the evaporation temperature increased with the inlet temperature of chilled water for considering the evaporator as a heat exchanger when the average temperature of the hot-fluid side rose under the constant mass flow rate and heat transfer characteristic. Therefore, the evaporation pressure increased along with the absorption pressure, which promoted the ability of the vapor absorption of a strong solution in the absorber, consequently reducing the mass fraction of the weak solution. In addition, the condensation temperature showed an upward trend with the chilled water inlet temperature rising, which resulted in the mass fraction of the strong solution decreasing because of the decline in the generator temperature and restriction of the Dühring rule in Equation (10). However, the concentration decrement of the strong solution was less than the weak solution, which caused the deflation ratio of the chiller to increase and the solution circulation ratio to decrease, therefore producing a greater quantity of refrigerant vapor in a cycle. This was the reason for the increase in the condenser and evaporator heat loads. For the absorber and generator, to generate the vapor or absorb the refrigerant accompanied with the phase change involved a considerable heat transfer in the process that was more than the same amount of liquid solution. Thus, an increasing of refrigerant quantity yielded to the heat loads in the absorber and generator augment as the mass flow rate of the solution pump remained constant.

Unlike the previous situations, in this case, the temperatures of the solution exiting from the absorber and generator presented an opposite trend according to the calculations, which showed an upward trend for the former and downward for the latter. The reason was that the temperatures of weak and strong solutions were closely related to the heat capacities of the absorber and generator, which mainly depended on the solution circulation ratio, but more importantly, the specific mass flow rate of each state point played a crucial role.

In both circumstances, the increased rate in cooling capacity was about 2–3% °C^{-1} with the inlet temperature of chilled water rising and ranged between 10 °C and 25 °C for the LiBr-H$_2$O system in comparison to the LiCl-H$_2$O system between 12 °C and 25 °C. The *COP* increased from 0.68 to 0.77 for the LiBr-H$_2$O system and from 0.72 to 0.79 for the LiCl-H$_2$O system, whereas the exergetic efficiency of both chillers presented a downward trend with an increasing decrement in the range of ≈5–22% °C^{-1}. In addition, the *COP* or exergetic efficiency gap between the two chillers shrunk.

In this section, the effects of a single change in the inlet temperature of the external circuit flows have been researched. From all of the above calculation results, something in common can be observed, which is that the temperatures of the fluid exiting from components of two chillers were almost the same as each other. This shows that the difference of heat duties in the same component of two chillers was mainly up to the thermodynamic properties of the solution, especially for the generator and absorber. That is why the LiCl-H$_2$O was superior to LiBr-H$_2$O in performance, but limited in application.

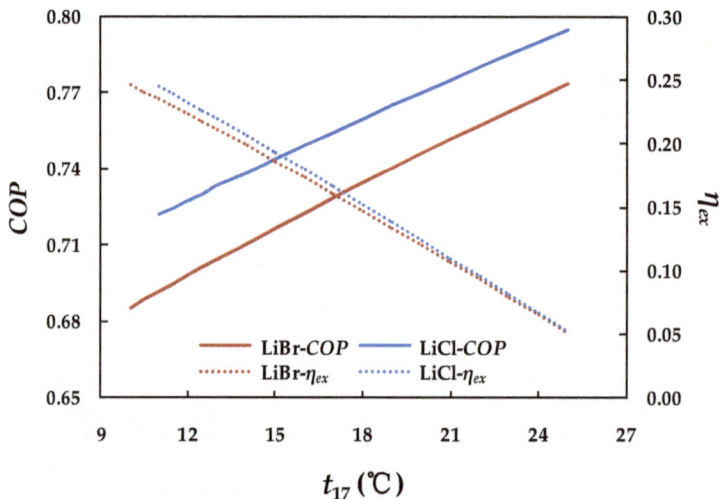

Figure 19. Variation of *COP* and exergy efficiency with the chilled water inlet temperature for the two absorption chillers.

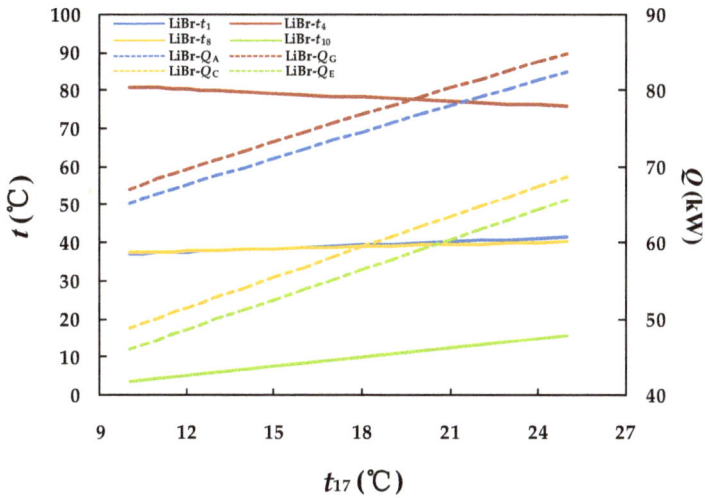

Figure 20. Variation of the exit temperature and heat capacity of each component with the chilled water inlet temperature for the LiBr-H$_2$O absorption chiller.

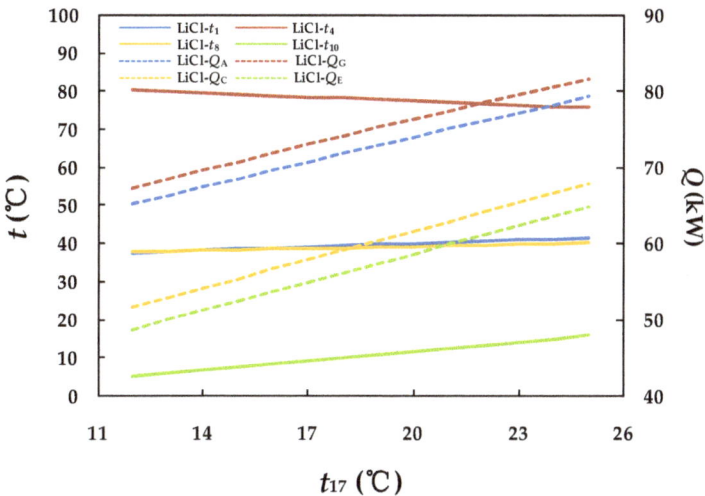

Figure 21. Variation of the exit temperature and heat capacity of each component with the cooling water inlet temperature for the LiCl-H$_2$O absorption chiller.

5. Conclusions

In this study, the use of LiBr-H$_2$O and LiCl-H$_2$O working pairs in the absorption refrigeration system were investigated from two aspects with multiple evaluation criteria. First, a novel calculation model was developed to investigate the performance of a single effect ARS and the model was validated with the available data in the literature. The effect of various design parameters was examined using energy and efficiency analyses. Second, the LiBr-H$_2$O and LiCl-H$_2$O absorption chillers with a capacity of 50 kW were designed with the same parameters. Then, the computer program of the absorption chiller was established to examine its performance under off-design external temperature conditions based on energy and exergy analysis. The main conclusions that can be drawn are as follows.

The *COP* and efficiency ratio in both systems increased with the generator temperature when the condenser and absorber temperatures were kept constant, but further increases brought down the efficiency ratio instead of being straight and level as for the *COP*. Lower temperatures of the condenser and absorber led to a higher *COP* and efficiency ratio, as well as their maximum values, consequently reducing the optimum generator temperature. The *COP* of both systems increased with evaporator temperature, but the growth rate weakened and all the values approached 0.8 eventually.

The variation of the generator temperature was limited as an upper bound by condenser temperature for the strong solution crystallization after the solution heat exchanger. The lowest generator temperature of the weak solution was decided by the absorber and evaporator temperatures. As a consequence, for better performance, a higher condensation or absorption temperature is more appropriate with a higher generator temperature, and vice versa for a lower evaporator temperature.

The studies on off-design conditions indicate that the inlet temperature of hot water with an optimum *COP* of the chiller was about 98 °C and 83 °C for LiBr-H_2O and LiCl-H_2O, respectively. However, the exergetic efficiency of the absorption chiller was negatively affected by the increase of the heat source temperature. In addition, the influences of the cooling water inlet temperature between the condenser and absorber on the performance were remarkable for both absorption chillers. The *COP* and exergetic efficiency degraded dramatically when the cooling water temperature was over the 41 °C for both chillers. With the inlet temperature of the chilled water rising, the cooling capacity of both chillers increased with the rate about 2–3% $°C^{-1}$ and the *COP* also increased, but this was opposite to the case of the exergetic efficiency.

The final conclusion of this study is that the LiCl-H_2O system has great limitations for practical application due to the crystallization problem though it performed better than a conventional LiBr-H_2O system under identical operating conditions. Compared to the LiBr-H_2O system, a LiCl-H_2O system is more appropriate for the situation where it is provided with a somewhat lower generator temperature or higher condenser temperature. After all, the diversity performances of the two working pairs depended on their own thermodynamic properties.

Supplementary Materials: The computer programs and result data are available online at http://doi.org/10.5281/zenodo.3251165.

Author Contributions: Formal analysis, N.G. and Y.Z.; Funding acquisition, Z.Q.; Investigation, Z.Q.; Methodology, J.R.; Project administration, Z.Q. and Z.Y.; Software, J.R.; Supervision, Z.Y.; Validation, J.R.; Writing—Original draft, J.R. and Y.Z.; Writing—Review and editing, Z.Y. and N.G.

Funding: This research was supported by National Natural Science Foundation of China (grant number 41276196 and 51676144) and China Scholarship Council (grant number 201706955097).

Acknowledgments: The authors gratefully acknowledge the assistance of lab members and also the reviewers for their insightful comments on the manuscript.

Nomenclature

Symbols		Subscripts and Superscripts	
a	Dühring gradient	*0*	reference point
a_0, a_1	constants	*A*	absorber
b	Dühring intercept	*a*	equilibrium state
b_0, b_1	constants	*C*	condenser
c_p	specific heat (kJ·kg^{-1}·K^{-1})	*c*	Carnot
H	energy flow rate (kW)	*co*	condensation
h	latent heat (kJ·kg^{-1})	*E*	evaporator
m	mass flow rate (kg·s^{-1})	*eff*	efficiency ratio
Q	heat load (kW)	*ev*	evaporation

t	temperature (°C)	*ex*	exergetic
W	power (kW)	G	generator
x	mass concentration of solution	K	Kelvin temperature scale
		k	process unit
		i	process stream
		j	system substance (water, LiBr, LiCl)
	Greek Letters	*in*	inlet
Δ	refers to the difference between two values	l	liquid
η	efficiency	*lm*	logarithmic mean temperature difference
ε	heat exchanger effectiveness	r	refrigerant
$\overline{X}$	process quantity	*out*	outlet
		SHE	solution heat exchanger
		s	dew point
		ss	strong solution
	Abbreviations	v	vapor
COP	coefficient of performance	u	external utility (cooling water, chilled water, and hot water)
CR	circulation ratio	w	water
UA	heat transfer characteristics (kW·K^{-1})	*ws*	weak solution

References

1. Ebrahimi, K.; Jones, G.F.; Fleischer, A.S. A review of data center cooling technology, operating conditions and the corresponding low-grade waste heat recovery opportunities. *Renew. Sustain. Energy Rev.* **2014**, *31*, 622–638. [CrossRef]

2. Mbikan, M.; Al-Shemmeri, T. Computational Model of a Biomass Driven Absorption Refrigeration System. *Energies* **2017**, *10*, 234. [CrossRef]

3. Alobaid, M.; Hughes, B.; Calautit, J.K.; O'Connor, D.; Heyes, A. A review of solar driven absorption cooling with photovoltaic thermal systems. *Renew. Sustain. Energy Rev.* **2017**, *76*, 728–742. [CrossRef]

4. Mahmoudi, S.; Kordlar, M.A. A new flexible geothermal based cogeneration system producing power and refrigeration. *Renew. Energy* **2018**, *123*, 499–512. [CrossRef]

5. Srikhirin, P.; Aphornratana, S.; Chungpaibulpatana, S. A review of absorption refrigeration technologies. *Renew. Sustain. Energy Rev.* **2001**, *5*, 343–372. [CrossRef]

6. Sun, J.; Fu, L.; Zhang, S. A review of working fluids of absorption cycles. *Renew. Sustain. Energy Rev.* **2012**, *16*, 1899–1906. [CrossRef]

7. Wang, J.; Yan, Z.; Wang, M.; Dai, Y. Thermodynamic analysis and optimization of an ammonia-water power system with LNG (liquefied natural gas) as its heat sink. *Energy* **2013**, *50*, 513–522. [CrossRef]

8. Günhan, T.; Ekren, O.; Demir, V.; Hepbasli, A.; Erek, A.; Şahin, A. Şencan Experimental exergetic performance evaluation of a novel solar assisted LiCl–H$_2$O absorption cooling system. *Energy Build.* **2014**, *68*, 138–146. [CrossRef]

9. Moreno-Quintanar, G.; Rivera, W.; Best, R. Comparison of the experimental evaluation of a solar intermittent refrigeration system for ice production operating with the mixtures NH$_3$/LiNO$_3$ and NH$_3$/LiNO$_3$/H$_2$O. *Renew. Energy* **2012**, *38*, 62–68. [CrossRef]

10. Cai, D.H.; Jiang, J.K.; He, G.G.; Li, K.Q.; Niu, L.J.; Xiao, R.X. Experimental evaluation on thermal performance of an air-cooled absorption refrigeration cycle with NH$_3$-LiNO$_3$ and NH$_3$-NaSCN refrigerant solutions. *Energy Conv. Manag.* **2016**, *120*, 32–43. [CrossRef]

11. Asfand, F.; Stiriba, Y.; Bourouis, M. Performance evaluation of membrane-based absorbers employing H$_2$O/(LiBr + LiI + LiNO$_3$ + LiCl) and H$_2$O/(LiNO$_3$+KNO$_3$+NaNO$_3$) as working pairs in absorption cooling systems. *Energy* **2016**, *115*, 781–790. [CrossRef]

12. Alvarez, M.E.; Esteve, X.; Bourouis, M. Performance analysis of a triple-effect absorption cooling cycle using aqueous (lithium, potassium, sodium) nitrate solution as a working pair. *Appl. Therm. Eng.* **2015**, *79*, 27–36. [CrossRef]

13. Gommed, K.; Grossman, G.; Ziegler, F. Experimental Investigation of a LiCl-Water Open Absorption System for Cooling and Dehumidification. *J. Sol. Energy Eng.* **2004**, *126*, 710–715. [CrossRef]

14. Conde, M.R. Properties of aqueous solutions of lithium and calcium chlorides: Formulations for use in air conditioning equipment design. *Int. J. Therm. Sci.* **2004**, *43*, 367–382. [CrossRef]

15. Kim, D.; Ferreira, C.I. A Gibbs energy equation for LiBr aqueous solutions. *Int. J. Refrig.* **2006**, *29*, 36–46. [CrossRef]

16. Pátek, J.; Klomfar, J. Solid–liquid phase equilibrium in the systems of LiBr–H₂O and LiCl–H₂O. *Fluid Phase Equilibria* **2006**, *250*, 138–149. [CrossRef]

17. Pátek, J.; Klomfar, J. A computationally effective formulation of the thermodynamic properties of LiBr–H₂O solutions from 273 to 500K over full composition range. *Int. J. Refrig.* **2006**, *29*, 566–578. [CrossRef]

18. Pátek, J.; Klomfar, J. Thermodynamic properties of the LiCl–H₂O system at vapor–liquid equilibrium from 273 K to 400 K. *Int. J. Refrig.* **2008**, *31*, 287–303. [CrossRef]

19. Parham, K.; Atikol, U.; Yari, M.; Agboola, O.P. Evaluation and optimization of single stage absorption chiller using (LiCl+H₂O) as the working pair. *Adv. Mech. Eng.* **2013**, *2013*. [CrossRef]

20. Patel, J.; Pandya, B.; Mudgal, A. Exergy Based Analysis of LiCl-H₂O Absorption Cooling System. *Energy Procedia* **2017**, *109*, 261–269. [CrossRef]

21. Bellos, E.; Tzivanidis, C.; Antonopoulos, K.A. Exergetic and energetic comparison of LiCl-H₂O and LiBr-H₂O working pairs in a solar absorption cooling system. *Energy Convers. Manag.* **2016**, *123*, 453–461. [CrossRef]

22. Gogoi, T.; Konwar, D. Exergy analysis of a H₂O–LiCl absorption refrigeration system with operating temperatures estimated through inverse analysis. *Energy Convers. Manag.* **2016**, *110*, 436–447. [CrossRef]

23. She, X.H.; Yin, Y.G.; Xu, M.F.; Zhang, X.S. A novel low-grade heat-driven absorption refrigeration system with LiCl-H₂O and LiBr-H₂O working pairs. *Int. J. Refrig.* **2015**, *58*, 219–234. [CrossRef]

24. Ochoa, A.; Dutra, J.; Henríquez, J.; Dos Santos, C. Dynamic study of a single effect absorption chiller using the pair LiBr/H₂O. *Energy Convers. Manag.* **2016**, *108*, 30–42. [CrossRef]

25. Ochoa, A.; Dutra, J.; Henríquez, J.; Dos Santos, C.; Rohatgi, J. The influence of the overall heat transfer coefficients in the dynamic behavior of a single effect absorption chiller using the pair LiBr/H₂O. *Energy Convers. Manag.* **2017**, *136*, 270–282. [CrossRef]

26. Marc, O.; Sinama, F.; Praene, J.-P.; Lucas, F.; Castaing-Lasvignottes, J. Dynamic modeling and experimental validation elements of a 30 kW LiBr/H₂O single effect absorption chiller for solar application. *Appl. Therm. Eng.* **2015**, *90*, 980–993. [CrossRef]

27. Kohlenbach, P.; Ziegler, F. A dynamic simulation model for transient absorption chiller performance. Part I: The model. *Int. J. Refrig.* **2008**, *31*, 217–225. [CrossRef]

28. Kohlenbach, P.; Ziegler, F. A dynamic simulation model for transient absorption chiller performance. Part II: Numerical results and experimental verification. *Int. J. Refrig.* **2008**, *31*, 226–233. [CrossRef]

29. Hellmann, H.M.; Schweigler, C.; Ziegler, F. A Simple Method for Modelling the Operating Characteristics of Absorption Chillers. In Proceedings of the Thermodynamics Heat and Mass Transfers of Refrigeration Machines and Heat Pumps Seminar Eurotherm No. 59, Nancy, France, 6–7 July 1998; pp. 219–226.

30. Arnavat, M.P.; López-Villada, J.; Bruno, J.C.; Coronas, A. Analysis and parameter identification for characteristic equations of single- and double-effect absorption chillers by means of multivariable regression. *Int. J. Refrig.* **2010**, *33*, 70–78. [CrossRef]

31. Gutiérrez-Urueta, G.; Rodríguez, P.; Ziegler, F.; Lecuona, A.; Rodríguez-Hidalgo, M. Extension of the characteristic equation to absorption chillers with adiabatic absorbers. *Int. J. Refrig.* **2012**, *35*, 709–718. [CrossRef]

32. Kim, D.; Ferreira, C.I. Analytic modelling of steady state single-effect absorption cycles. *Int. J. Refrig.* **2008**, *31*, 1012–1020. [CrossRef]

33. Wagner, W.; Cooper, J.R.; Dittmann, A.; Kijima, J.; Kretzschmar, H.J.; Kruse, A.; Mareš, R.; Oguchi, K.; Sato, H.; Stöcker, I.; et al. The IAPWS Industrial Formulation 1997 for the Thermodynamic Properties of Water and Steam. *J. Eng. Gas Turbines Power* **2000**, *122*, 150–184. [CrossRef]

34. Patel, H.A.; Patel, L.; Jani, D.; Christian, A. Energetic Analysis of Single Stage Lithium Bromide Water Absorption Refrigeration System. *Procedia Technol.* **2016**, *23*, 488–495. [CrossRef]

35. Edgar, T.F.; Himmelblau, D.M.; Lasdon, L.S. *Optimization of Chemical Processes*; McGraw-Hill: New York, NY, USA, 2001.

36. Gebreslassie, B.H.; Groll, E.A.; Garimella, S.V. Multi-objective optimization of sustainable single-effect water/Lithium Bromide absorption cycle. *Renew. Energy* **2012**, *46*, 100–110. [CrossRef]

37. Martínez, J.C.; Martinez, P.; Bujedo, L.A. Development and experimental validation of a simulation model to reproduce the performance of a 17.6 kW LiBr–water absorption chiller. *Renew. Energy* **2016**, *86*, 473–482. [CrossRef]

38. Kerme, E.D.; Chafidz, A.; Agboola, O.P.; Orfi, J.; Fakeeha, A.H.; Al-Fatesh, A.S. Energetic and exergetic analysis of solar-powered lithium bromide-water absorption cooling system. *J. Clean. Prod.* **2017**, *151*, 60–73. [CrossRef]

39. Kaita, Y. Simulation results of triple-effect absorption cycles. *Int. J. Refrig.* **2002**, *25*, 999–1007. [CrossRef]

40. Anand, D.; Kumar, B. Absorption machine irreversibility using new entropy calculations. *Sol. Energy* **1987**, *39*, 243–256. [CrossRef]

41. Kaushik, S.; Arora, A. Energy and exergy analysis of single effect and series flow double effect water–lithium bromide absorption refrigeration systems. *Int. J. Refrig.* **2009**, *32*, 1247–1258. [CrossRef]

energies

MDPI

Article

Modeling of a PCM TES Tank Used as an Alternative Heat Sink for a Water Chiller. Analysis of Performance and Energy Savings

Antonio Real-Fernández [1,*], Joaquín Navarro-Esbrí [2], Adrián Mota-Babiloni [2], Ángel Barragán-Cervera [2], Luis Domenech [1], Fernando Sánchez [1], Angelo Maiorino [3] and Ciro Aprea [3]

[1] Department of Mathematics, Universidad Cardenal Herrera-CEU, CEU Universities, Physics and Technological Sciences. C/San Bartolomé, 55, 46115 Alfara del Patriarca, Valencia, Spain; luis.domenech@uchceu.es (L.D.); fernando.sanchez@uchceu.es (F.S.)
[2] ISTENER Research Group, Department of Mechanical Engineering and Construction, Universitat Jaume I, Campus de Riu Sec s/n, E12071 Castelló de la Plana, Spain; navarroj@emc.uji.es (J.N.-E.); mota@uji.es (A.M.B.); abarraga@emc.uji.es (Á.B.C.)
[3] Department of Industrial Engineering, University of Salerno, Via Giovanni Paolo II 132, 84084 Fisciano (SA), Italy; amaiorino@unisa.it (A.M.); aprea@unisa.it (C.A.)
* Correspondence: antonio.real@uchceu.es

Received: 30 July 2019; Accepted: 19 September 2019; Published: 24 September 2019

Abstract: Phase change materials (PCMs) can be used in refrigeration systems to redistribute the thermal load. The main advantages of the overall system are a more stable energy performance, energy savings, and the use of the off-peak electric tariff. This paper proposes, models, tests, and analyzes an experimental water vapor compression chiller connected to a PCM thermal energy storage (TES) tank that acts as an alternative heat sink. First, the transient model of the chiller-PCM system is proposed and validated through experimental data directly measured from a test bench where the PCM TES tank is connected to a vapor compression-based chiller. A maximum deviation of 1.2 °C has been obtained between the numerical and experimental values of the PCM tank water outlet temperature. Then, the validated chiller-PCM system model is used to quantify (using the coefficient of performance, COP) and to analyze its energy performance and its dependence on the ambient temperature. Moreover, electrical energy saving curves are calculated for different ambient temperature profiles, reaching values between 5% and 15% taking the experimental system without PCM as a baseline. Finally, the COP of the chiller-PCM system is calculated for different temperatures and use scenarios, and it is compared with the COP of a conventional aerothermal chiller to determine the switch ambient temperature values for which the former provides energy savings over the latter.

Keywords: phase change material; thermal energy storage; vapor compression system; HVAC; energy efficiency

---

1. Introduction

In 1997 the existence of global warming was recognized worldwide through the Kyoto Protocol [1], resulting in a global commitment to reduce greenhouse gas emissions in which the anthropogenic CO_2 emissions were pointed out as the main cause. Later, in December 2015, the Paris Agreement [2] established measures to hold the increase in the global average temperature by 2100 to well below 2 °C above pre-industrial levels.

The International Institute of Refrigeration estimates that the total number of refrigeration, air-conditioning, and heat pump systems in operation worldwide is roughly 3 billion. This sector consumes about 17% of the overall used electricity worldwide [3] and accounts for 7.8% of global greenhouse gas emissions [4]. According to the International Energy Agency, the residential and commercial sectors were responsible for 43% of natural gas and 49% of world electricity final consumption in 2015 [5].

Phase change materials (PCMs) are widely used to reduce energy consumption in buildings. They can be incorporated as a passive element in buildings to increase the thermal inertia of its components. Applying PCM to the building envelope can reduce indoor temperature peaks and heating, ventilation and air-conditioning (HVAC) cooling load [6]. PCMs can also be utilized to save energy in radiant floor heating systems [7] or included in the roof to reduce the through roof heat gain in buildings [8].

Another application of PCM to enhance energy efficiency is its use in thermal energy storage (TES) systems to provide a greater energy storage capacity in a narrower temperature range. In this way, Anisur et al. [9] highlighted that TES, in general and particularly PCMs, could provide energy savings and, hence, reductions in CO_2 emissions. PCM TES systems can be included in solar water heating systems to provide higher heat storage and thermal efficiency [10]. They can also be included in an air-PCM heat exchanger for passive cooling of buildings where a cooling storage unit is cooled during the night to reduce the indoor air temperature during the day [11]. Ventilation systems with PCM can also be applied to air-conditioned buildings granting electricity energy savings [12].

Vapor compression systems (VCS) with aerothermal dissipation are an extended solution for HVAC systems in residential and commercial buildings [13]. PCM based TES systems can improve their efficiency with a load shift from on-peak to off-peak periods to take advantage of a lower electricity tariff and a lower condensing air temperature [14]. Ruddell et al. [15] concluded that, in a city with high air-conditioning peak demand (Phoenix, Arizona, USA), the distributed cold TES technology could reduce the peak air-conditioning electrical power demand by 23% and the peak total electrical power demand, by 13%.

However, as a disadvantage, energy storage for cooling purposes must be done at temperatures widely below ambient temperature, typically between 0 and 10 °C [16]. Therefore, all PCM TES systems experience tank thermal losses that typically range between 1% and 5% per day [17]. These losses make cold storage strategies inefficient when the energy is stored for long periods.

A less extended application of TES systems consists in using PCMs as an alternative heat sink to improve the performance of the system. During the day, when the outside temperature rises and lowers the energy performance of the chiller, the PCM tank provides a lower temperature heat sink. Then, the heat accumulated in the tank is released at night when the temperature falls. These applications use PCMs with melting temperatures around the average outdoor temperature, which reduces the ambient thermal losses. Until today, few publications have assessed this topic. Zhao and Tan analysed the use of a finned shell-and-tube PCM thermal storage system to dissipate the heat of a prototype thermoelectric cooling system [18] and an air-conditioning application [19]. The authors focused on the design parameters of the PCM TES and how they affected the heat exchange on it.

There is a lack of research on the parameters that affect the energy performance of vapor compression chillers when a PCM TES is used as a heat sink. Making a direct COP comparison between them and a conventional chiller can be considered difficult. The electrical consumption of the first system is divided into two stages, differed in time (day and night), and the temperatures of the TES system and the ambient air influence its energy performance. Therefore, this work models and studies the use of an experimental PCM TES system as an alternative dissipation sink for a chiller. It focuses on the system energy performance and savings and analyses how different ambient temperature profiles affect it. The model is defined and appropriately validated using experimental data from a test rig assembled and tested at different operating conditions. It is complemented by a methodology that evaluates the energy savings provided by this kind of chiller-PCM system. After that, this model is used to perform simulations of each functioning mode of the system that represent different ambient temperature conditions using energy performance curves, which are included in this paper. Finally, the energy performance of the chiller-PCM mode is compared with that of a conventional aerothermal mode to determine the best temperature to switch between them.

2. Chiller-PCM System Description

A schematic diagram of the proposed chiller-PCM system is presented in Figure 1. The chiller is connected to the cooled room (cold sink) through the evaporator and to the two alternative heat sinks, a fan, and the PCM tank, through the condenser. During the daytime, at low ambient temperatures, the chiller uses the fan as a sink (aerothermal mode). When the ambient temperature rises and lowers the COP of the chiller, the heat sink is switched to the PCM tank obtaining a lower condensing temperature (PCM discharge mode). During the nighttime, when the ambient temperature falls, the heat stored in the PCM tank is released using the fan (PCM charge mode). The PCM charge and discharge modes (PCM mode) used instead of the aerothermal mode can imply a reduction in the electrical consumption of the overall system.

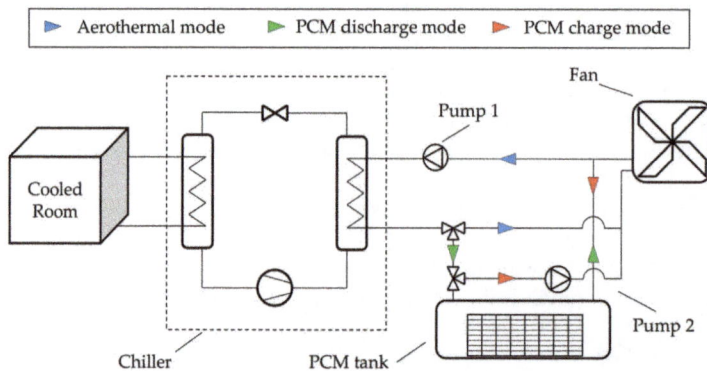

Figure 1. Schematic diagram of the chiller-phase change materials (PCM) system.

A monitored test bench has been adapted to characterize the performance of the described chiller-PCM system. This experimental facility consisted of a PCM tank connected to a VCS with a proportional integral derivative (PID) controlled thermal load circuit on the evaporator side and to an aerothermal dissipation system.

The PCM tank was made of polymeric material, and its dimensions were $3 \times 0.75 \times 0.75$ meters. Inside it, 192 PCM containers were arranged in 16 layers of 12 containers each (Figure 2). The PCM containers were placed in the center of the tank, leaving 0.5 meters free on each side for a proper water flow homogenization. Furthermore, water was driven to and from the tank using PVC T-shaped pipes with multiple holes for proper water distribution.

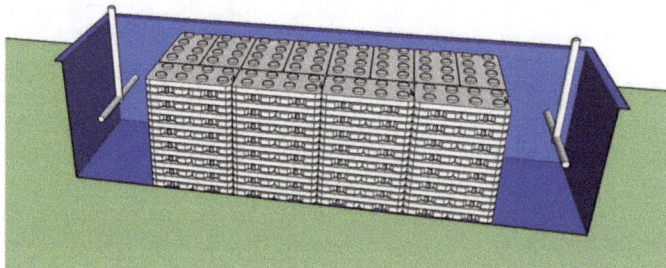

Figure 2. PCM storage tank 3D representation.

The overall latent heat of the PCM containers set was 1674 MJ, sized to dissipate a 6-kW thermal load over more than 7 hours. The PCM used was the commercially available salt hydrate S27 from PCM Products Ltd [20], and its main properties are shown in Table 1, where latent heat was obtained from the study carried out by Barreneche et al. [21]. Moreover, S27 was encapsulated in high-density polyethylene 0.5 × 0.25 × 0.032 m containers.

Table 1. Salt hydrate S27 main properties.

Parameter	Value
Melting temperature (°C)	27
Density (kg·m^{-3})	1530
Specific latent heat (kJ·kg^{-1})	150
Specific heat (kJ·kg^{-1}·K^{-1})	2.20
Thermal conductivity (W·m^{-1}·K^{-1})	0.54

The main components of the VCS were a shell (refrigerant) and tube (water) condenser, an R134a thermostatic expansion valve, a flat plate evaporator (glycol based secondary fluid), and an open alternative compressor driven by a 7.5 kW electric motor. The working fluid was the hydrofluorocarbon R134a, typically used in water chillers. The main temperatures and pressures of the system were registered as well as the refrigerant mass flow rate, secondary fluids volumetric flow rate and compressor electricity consumption. In the thermal load circuit, 15 kW resistances controlled by a PID set the temperature of a thermally insulated deposit. The aerothermal dissipation system consisted of a 190 W fan and a 7.9 kW chiller.

3. Model Description and Validation

The model of the chiller-PCM system has been designed as a model of the PCM tank coupled to a model of the VCS, while the water pumps and the fan have been modeled using datasheet characteristics. Both models have been programmed using MATLAB 8.

A model of the VCS has been designed to calculate its main energy parameters from the condensing conditions. Therefore, the evaporator conditions have been set constant to obtain chilled water operating temperature of 7/12 °C. The configuration of the compressor and the expansion valve have been set constant too. Thereby, the only parameters defining the performance of the VCS are the condenser conditions, namely the condenser water inlet temperature ($T_{cnd,in}$) and its volumetric flow rate ($\dot{V}$) (Figure 3). A regression model has been obtained through a set of measurements on the experimental facility. The adjusted R^2 coefficients are shown in Table 2.

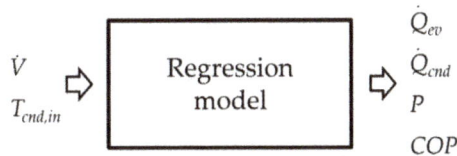

Figure 3. Vapor compression systems (VCS) model structure.

Table 2. Adjusted R^2 coefficients for the Vapor compression systems (VCS) regression model.

Parameter	Range	R^2 Coefficient
$\dot{Q}_{cnd}$ (kW)	[7.37–9.28]	0.9558
$\dot{Q}_{ev}$ (kW)	[5.70–7.71]	0.9738
P (kW)	[2.33–3.00]	0.9912
COP	[1.91–3.31]	0.9958

The model of the PCM tank has been designed to provide the water outlet temperature ($T_{w,out}$) given the water volumetric flow rate ($\dot{V}$), the heat transfer rate in the condenser ($\dot{Q}_{cnd}$), the state of the PCM in a specific instant (i) and the water outlet temperature obtained in the previous instant.

The state of the PCM in a specific instant is expressed through its temperature ($T_{PCM,i}$). Finally, the model should provide the state of the PCM for the next iteration (*i+1*) (Figure 4).

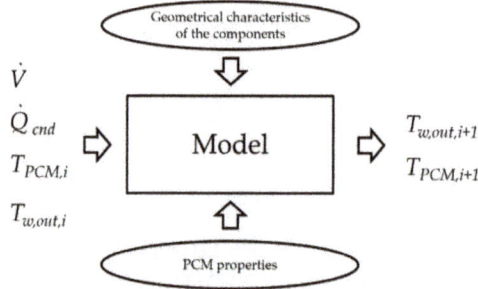

Figure 4. PCM storage tank model structure.

The water inlet temperature ($T_{w,in}$) is calculated using the input parameters $\dot{Q}_{cnd}$ and $\dot{V}$ and the outlet temperature obtained in the previous iteration:

$$T_{w,in,i+1} = T_{w,out,i} + \frac{\dot{Q}_{cnd}}{\dot{V}\,\rho_w\,C_{pw}} \tag{1}$$

The tank is divided into two homogenization zones at the inlet and the outlet of the water flow, being PCM containers stacked in the geometrical center of it. For calculation purposes, in the model, the PCM has been discretized. It is considered divided into eight sections with the width of the container (Figure 5). First, a perfect mixture is considered between the water inlet and the first homogenization zone. Then, the water enters the first PCM section at the temperature of the first homogenization zone (T_{hz1}), successively exchanging heat with each section. After that, the water, at the outlet temperature of the last PCM section ($T_{w,sc8,out}$), enters the second homogenization zone where again a perfect mixture is considered. Finally, the water leaves the tank at the temperature of the second homogenization zone (T_{hz2}).

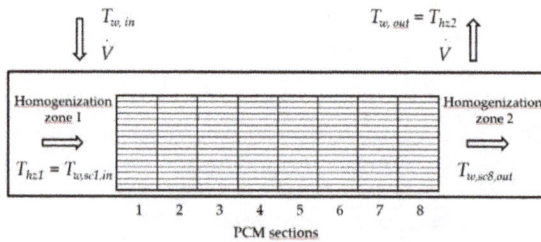

Figure 5. Distribution of the PCM storage tank as considered in the model.

Each PCM section can be considered as a heat exchanger with a thermal resistance between the water and the PCM (Figure 6). The inlet temperature of each section is the outlet temperature of the previous one. The thermal resistance between the water and the PCM (R_{w-PCM}) consists of a convection resistance between the water and the container and a conduction resistance in the container itself as expressed in Equation (2). Furthermore, when the PCM begins to melt, the liquid phase creates an additional natural convection resistance between the container and the solid PCM as formulated in Equation (3).

$$R_{w-PCM} = R_{w-cont} + R_{cont} \tag{2}$$

$$R_{w-PCM} = R_{w-cont} + R_{cont} + R_{cont-PCM} \tag{3}$$

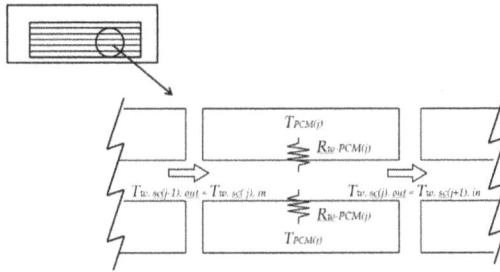

Figure 6. Distribution of the PCM storage tank as considered in the model.

The convection thermal resistance between the water and the container (R_{w-cont}) is calculated from the film coefficient (h) and the contact surface (S) which is obtained from the Nusselt number (Nu):

$$R = \frac{1}{h\,S}, \tag{4}$$

$$h = \frac{Nu\,k}{d_h} \tag{5}$$

The Nusselt number is calculated using the Dittus Boelter correlation [22], where k is equal to 0.4 when heating the fluid and to 0.3 when cooling it.

$$Nu = 0.023\,Re^{0.8}\,Pr^k \tag{6}$$

The container resistance (R_{cont}) is a conventional conduction one, and that of the melted PCM between the container and the solid PCM ($R_{cont\text{-}PCM}$) is calculated considering the generic natural convection equation:

$$Nu = cRa^n \tag{7}$$

where the parameters c and n depend on the geometry of the container and have been adjusted using the experimental measurements obtaining a value of 0.45 and 0.105, respectively.

The heat transfer rate of each PCM section is calculated applying the NTU-effectiveness method to the thermal resistance, and then, the outlet water temperature of the considered section can be obtained.

$$T_{w,sc(j),out} = T_{w,sc(j),in} - \frac{\dot{Q}_j}{\dot{V}\,\rho_w\,Cp_w} \tag{8}$$

The latent heat of the PCM is modeled as a Dirac delta function (δ) overlapping the specific heat as proposed by J. Ning-Wei [23].

$$\delta(T) = \frac{e^{-\left(\frac{T-T_m}{b^2}\right)}}{\sqrt{\pi}b} \tag{9}$$

$$Cp(T) = H(T_{melt} - T)\,Cp_{sol} + \delta(T)\,L + H(T - T_{melt})\,Cp_{liq} \tag{10}$$

where b is a parameter that defines the width of the Dirac delta function, and the function $H(x)$ takes the following values:

$$H(x) = \begin{cases} 0 & x < 0 \\ 1 & x > 0 \end{cases} \tag{11}$$

The PCM temperature for the next iteration ($i+1$) for the section j can be calculated with the heat transfer rate and the specific heat of the PCM.

$$T_{PCM(j),i+1} = T_{PCM(j),i} + \frac{\int_{i}^{i+1} \dot{Q}_j}{m_{PCM(j)} \, Cp_{PCM(j)}} \tag{12}$$

Finally, the liquid fraction of the PCM can be calculated as the integral of the Dirac delta function.

$$f = \int_{-\infty}^{T} \delta(T)dT \tag{13}$$

To validate the model of the PCM TES tank a set of experimental measurements have been carried out using the facility described in Section 2. These measurements have also been simulated using the model. The output parameters of both experimental and model results have been compared. Four different discharge mode measurements have been arranged at four different volumetric flow rates (0.70, 0.95, 1.20, and 1.45 $m^3 \cdot h^{-1}$). For the charge mode, two measurements have been arranged at two different water flow rates (0.95 and 1.45 $m^3 \cdot h^{-1}$).

Figure 7 shows the performance of the tank water outlet temperature during a typical discharge (a) and charge (b) test. As can be observed, the model brings a good representation of the PCM tank thermal performance, especially during the phase change of the PCM. In the discharge mode, the maximum deviation obtained between the numerical and experimental values of the PCM TES tank water outlet temperature was of 1.2 °C. In the charge mode, a maximum deviation of 1.1 °C was obtained between numerical and experimental values of $T_{w,out}$. The maximum deviation was observed usually at the beginning and the end of the simulation, being the difference during the phase change smaller.

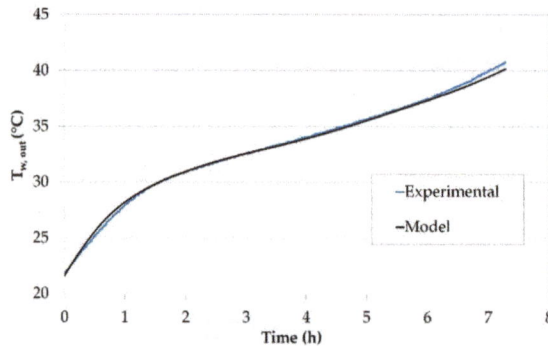

(a)

Figure 7. *Cont.*

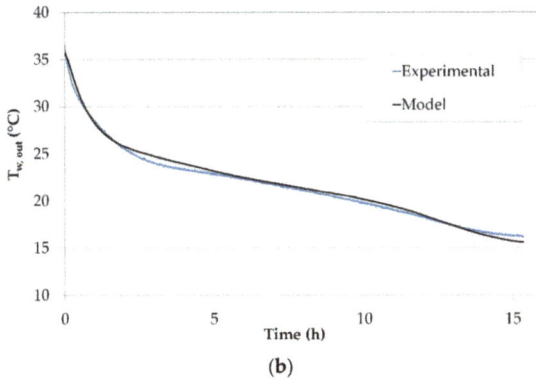

Figure 7. Example of the tank model validation for (a) a discharge test ($\dot{V} = 1.20$ m$^3 \cdot$h^{-1}) and for (b) a charge test ($\dot{V} = 0.95$ m$^3 \cdot$h^{-1}).

It must be remarked that, although the PCM melts at 27 °C, the values of $T_{w,out}$ during the discharge test rise over this value with an ascending slope. This temperature difference depends on the thermal conductivity between the PCM and the water in the tank for the selected products and configuration.

4. Consumption Analysis

To evaluate the performance of the proposed chiller-PCM system, first, the consumption of the three different operating modes was studied separately. Then the consumption of the PCM charge and discharge modes were added to compare the PCM mode with the aerothermal mode.

4.1. Operating Modes and Associated Consumptions

In the aerothermal mode, the chiller cools the conditioned space rejecting the heat to the air using the fan. The three components that consume electrical energy during the operation in this mode are the compressor of the chiller, the pump 1, and the fan.

$$W_{aero} = \int_0^{t_1} (P_{comp} + P_{fan} + P_{p1})dt = W_{comp} + W_{fan} + W_{p1} \tag{14}$$

The two remaining operating modes correspond to the use of the PCM tank. First, in the PCM discharge mode, the chiller cools the conditioned space storing the condenser's heat in the PCM tank. The two components that consume electrical energy in this mode are the compressor of the chiller and the pump 1.

$$W_{PCM-dch} = \int_0^{t_1} (P_{comp} + P_{p1})dt = W_{comp} + W_{p1} \tag{15}$$

Then, in the PCM charge mode, the heat stored in the PCM tank is rejected to the air using the fan. The two components that consume electrical energy in this mode are the fan and the pump 2.

$$W_{PCM-ch} = \int_0^{t_2} (P_{fan} + P_{p2})dt = W_{fan} + W_{p2} \tag{16}$$

The total energy consumed in the PCM mode to extract an amount of heat Q_{ev} from the cold sink is the sum of the energy consumed during the PCM charge and discharge modes.

$$W_{PCM} = W_{PCM-dch} + W_{PCM-ch} = W_{comp} + W_{p1} + W_{fan} + W_{p2} \tag{17}$$

It is important to remark that the charging process is differed in time from the cold generation and has a different duration from the discharge process. As the electrical consumption during the charge and discharge processes depends on the temperature of the tank, it is important to assign the corresponding charge electrical consumption to each use of the chiller. Thus, for a Q_{ev} generated that causes an increase in the liquid fraction of the PCM from f_1 to f_2, the charge electrical consumption associated to this Q_{ev} is the one necessary to reduce the liquid fraction of the PCM from f_2 to f_1.

4.2. Consumption Analysis of The Different Operating Modes

The most commonly used parameter to evaluate the performance of a chiller is the COP. However, it presents two limitations to be used in the performance of the PCM charge and discharge modes. First, the COP of the charge and discharge modes would not provide a real approximation, as it would only include a fraction of the PCM mode electrical consumption. Second, the COP of both charge and discharge modes cannot be added to the calculation of the global COP of the PCM mode. Therefore, the consumption ratio (CR) was used to evaluate the performance of the PCM charge and discharge modes. It is defined as the ratio of the electrical consumption (input work) to the heat removed from the cold sink as expressed in Equation (18). The total CR of the PCM mode can be obtained as the addition of the PCM charge and discharge CRs, as given in Equation (19), and finally, the COP of the PCM mode is obtained with the Equation (20) as the inverse of R_{PCM}.

$$CR = \frac{W}{Q_{ev}} \tag{18}$$

$$CR_{PCM} = CR_{PCM-dch} + CR_{PCM-ch} \tag{19}$$

$$COP_{PCM} = \frac{1}{CR_{PCM}} \tag{20}$$

4.2.1. Consumption Analysis of The Aerothermal Mode

In the aerothermal mode, CR depends on the ambient temperature where the fan dissipates the heat. The higher the ambient temperature, the higher the value of CR (Figure 8). The values of CR_{aero} were used later to calculate the energy savings provided by the PCM mode. At higher ambient temperatures, the electric consumption increased, and the energy savings provided by the PCM mode increased likewise.

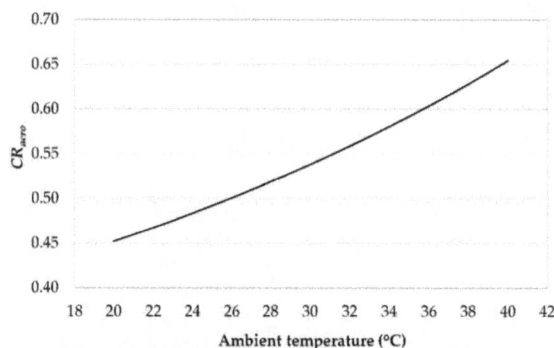

Figure 8. Value of the consumption ratio (CR) in the aerothermal mode for different ambient temperatures.

4.2.2. Consumption Analysis of The PCM Discharge Mode

In the PCM discharge mode, the performance of the chiller was independent of the ambient temperature but becomes dependent on the PCM tank temperature, which determined the condensing temperature of the chiller. To evaluate the *CR* of this mode, a simulation was carried out with the initial temperature of the PCM tank at 18 °C. The chiller dissipated the condenser heat in the PCM tank, increasing its temperature and thus, the value of *CR*, as shown in Figure 9a. Figure 9b shows the evaluation of this consumption ratio against the liquid fraction of the PCM, which is useful to link the electrical consumption of the PCM charge and discharge modes. Both figures show that, as the PCM melted, the temperature of the PCM tank increased and raised the value of *CR*. In addition, they highlight that for values of *f* close to the unity, the value of *CR* sharply increased.

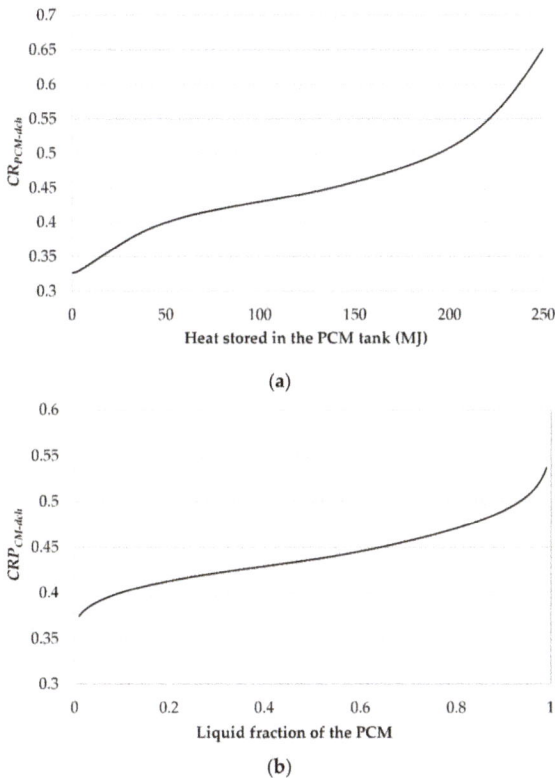

Figure 9. Value of *CR* in the PCM discharge mode against (a) heat stored in the tank and (b) liquid fraction of the PCM.

4.2.3. Consumption Analysis of The PCM Charge Mode

During the PCM charge mode, the fan dissipated the heat stored in the PCM tank to the ambient air. Consequently, the efficiency of this process depends on the ambient temperature. Additionally, as the tank lowers its temperature and the temperature difference between the air and the tank narrows the value of *CR* increased as represented in Figure 10a. Figure 10b shows how at higher ambient temperatures, the energy cost of solidifying the PCM to low values of *f* increased and, in some cases, it could be inefficient to completely solidify the PCM.

(a)

(b)

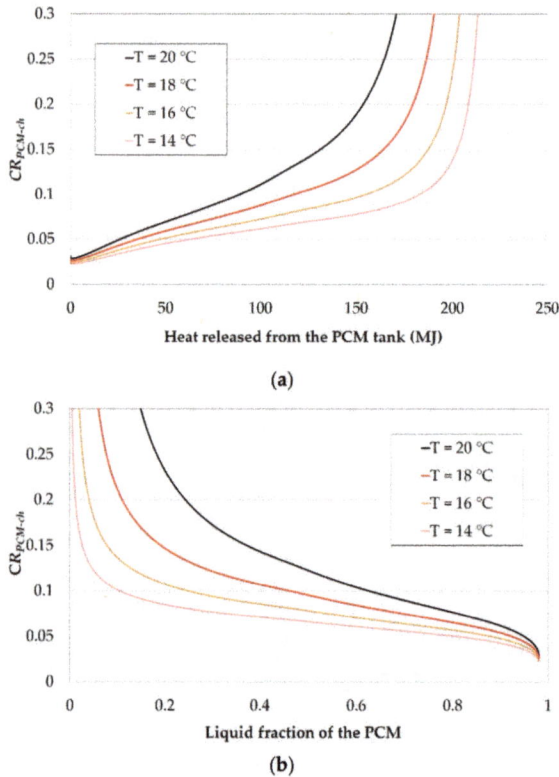

Figure 10. Value of *CR* in the PCM charge mode against (**a**) heat released from the tank and (**b**) liquid fraction of the PCM.

4.2.4. Consumption Analysis of The PCM Mode

The total electric consumption of the PCM mode is the sum of the charge and discharge modes. As the charge electric consumption depends on the f of the PCM, it must be evaluated at the same f than the discharge that caused it. Therefore, the charge and discharge values of *CR* are linked by the f of the PCM as the sum of Figures 9b and 10b. Figure 11a represents the *CR* of the PCM mode, which depends on the liquid fraction of the PCM and the temperature of the air during the charge of the PCM tank.

Although the discharge electric consumption was lower for low values of f, the higher electric consumption that required a complete solidification of the PCM increased the *CR* of the PCM mode, especially at higher night temperatures. At higher values of f, the charge electric consumption was low, but the higher discharge electric consumption slightly increased the CR_{PCM}. The lowest values of the consumption ratio take place for the central values of f.

Once the charge and discharge consumption ratios are unified, the COP of the system can be calculated using Equation (20). Again, the COP was maximized in the central values of f, where the lowest values of the consumption ratio were observed, as shown in Figure 11b.

(a)

(b)

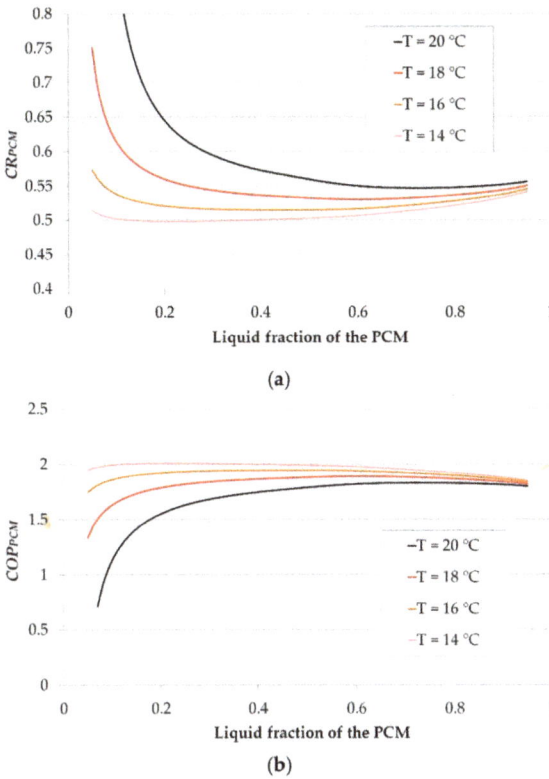

Figure 11. (a) CR and (b) coefficient of performance (COP) against liquid fraction of the PCM in the PCM mode.

4.3. Consumption Comparison Between Aerothermal and PCM Modes

The CR of the PCM operating mode was compared with the CR of the aerothermal mode to evaluate the convenience of connecting the PCM tank to the condenser of the chiller. The percentage of energy savings was the parameter selected for this comparison, and it is calculated as

$$Savings(\%) = \frac{CR_{aero} - CR_{PCM}}{CR_{aero}} \cdot 100 \tag{21}$$

CR_{aero} depends on the ambient temperature during the cold generation, whereas CR_{PCM} depends on the ambient temperature during the charge mode and the f of the PCM during both, the charge and discharge modes. Therefore, the energy savings must be evaluated given the ambient temperature conditions. Daily temperature can be represented as a sinusoidal curve with good accuracy [24] given a mean temperature T_m and an amplitude A defined as the difference between the maximum or minimum temperature and T_m.

First, an evaluation of the energy savings provided by the PCM mode was carried out considering different values of T_m and A. This analysis can guide for the sizing of the chiller-PCM system and the selection of the melting temperature of the PCM. Then the performance of the PCM mode was compared with the aerothermal mode for different use and temperature conditions. This analysis can determine the switch temperature above which the first mode provides energy savings over the second one.

4.3.1. Analysis of Energy Savings for Different Values of T_m

The relation between the T_m and the melting temperature of the PCM will determine the energy savings provided by the chiller-PCM system. A high value of T_m entails a high ambient temperature during the cold generation and the charge of the PCM tank and vice versa. For high values of T_m, the consumption ratio of the aerothermal mode will rise, whereas the cost of the PCM discharge mode remains constant, increasing the energy savings provided by the chiller-PCM system during the cold generation. However, a high T_m also entails a high ambient temperature during the PCM charge mode, increasing its consumption ratio and that of the overall PCM mode. However, for low values of T_m, the energy savings during the cold generation decrease, but the cost of the PCM charge mode will decrease as well.

To evaluate the effect of T_m on energy savings, a simulation was performed with four different values of T_m. Additionally, an arbitrary value of 8 °C was set for A. The resulting day and night temperatures are shown in Table 3. Day temperature was considered constant during the aerothermal mode and night temperature was considered constant during the PCM charge mode.

Table 3. Day and night temperatures for different values of daily mean temperature (T_m) and a daily temperature amplitude (A) of 8 °C.

T_m (°C).	Day Temperature (T_m+8) (°C)	Night Temperature (T_m-8) (°C)
23	31	15
25	33	17
27	35	19
29	37	21

The energy savings obtained for each T_m are represented in Figure 12. For high values of T_m concerning the melting temperature of the PCM high energy savings were obtained for values of f close to 1. However, the high cost of charging the PCM with a high night temperature makes the complete solidification of the PCM inefficient and part of its latent heat would remain unused. For low values of T_m, the charge mode had a lower consumption and energy savings were obtained for low values of f, but the energy savings were reduced for values of f close to 1. Considering the PCM and the chiller analyzed in this work, the best energy savings profile was obtained for a T_m of 25 °C, two degrees Celsius below the melting point of the PCM.

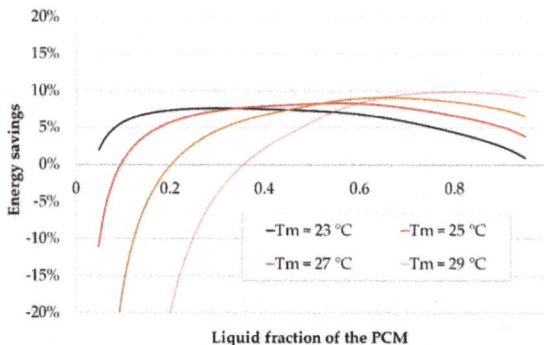

Figure 12. Energy savings obtained with the PCM mode for different values of daily mean temperature (T_m).

4.3.2. Analysis of Energy Savings for Different Values of A

A high amplitude in the ambient temperature profile entails a high day temperature and a low night temperature, and both result in higher energy savings. It is evident that the higher the A, the higher the energy savings. However, there is worth in determining the profile of energy savings for different values of A and the threshold for this parameter below which the energy savings disappear. Therefore, a simulation was designed for five different values of A. A constant value of 25 °C was chosen for T_m as it showed the best performance for the system. Table 4 shows the night and day temperatures for this simulation. Day temperature was considered constant during the aerothermal mode and night temperature was considered constant during the PCM charge mode.

Table 4. Day and night temperatures for different values of A and a T_m of 25 °C.

A (°C)	Day Temperature (25+A) (°C)	Night Temperature (25−A) (°C)
6	31	19
7	32	18
8	33	17
9	34	16
10	35	15

Simulation results show the expected increase in energy savings when A is increased (Figure 13). For a value of A equal to 6 °C and below the energy savings were negligible for the PCM, and the chiller analyzed in this work. A configuration with a lower thermal resistance between the PCM and the water would provide lower water temperatures at the outlet of the PCM tank and, thus, higher energy savings lowering the threshold value of A obtained in this simulation.

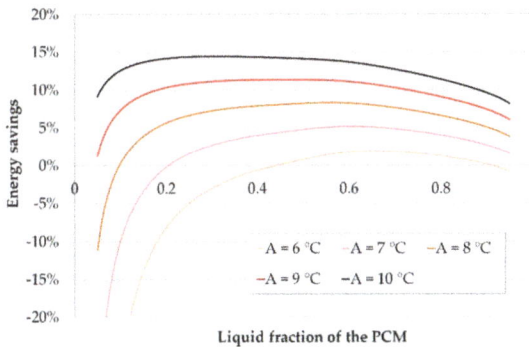

Figure 13. Energy savings obtained with the PCM mode for different values of daily temperature amplitude (A).

4.3.3. Performance Analysis for Different Use and Temperature Conditions

The energy savings obtained when replacing a conventional aerothermal chiller with the proposed chiller-PCM system depend on the ambient temperature during the cold generation (day) and the PCM charge mode (night) and on the liquid fraction of the PCM. As seen in the previous sections, there was a decrease in the energy savings provided by the PCM mode when completely melting or solidifying the PCM, whereas they reached their maximum value when f was near 0.5.

If the PCM discharge mode is used for a short period, the most energy convenient point of functioning can be selected, with an f value near 0.5. However, for longer periods, a larger range of f must be used, extending it to values with lower energy savings.

A simulation was performed comparing the COP of the aerothermal mode with that of the PCM mode, depending on the day ambient temperature. Three different values of the night ambient temperature (16, 18, and 20 °C) and three different use periods (1, 3, and 5 hours) were considered resulting in nine different scenarios. For periods of 1 hour, the most convenient fraction of f was selected, resulting in a high COP. For periods of 3 and 5 hours, the f fraction was extended to less convenient fractions reducing the global COP of the whole period.

Figure 14 shows the results of the simulation, where the intersection between the aerothermal COP and each of the nine PCM scenarios determined the switch day ambient temperature above which the PCM mode grants energy savings over the aerothermal mode. As can be observed, the scenario with the highest COP (use period of 1 hour and night temperature of 16 °C) provided energy savings for day temperatures above 22 °C while the scenario with the lowest COP (use period of 5 hours and night temperature of 20 °C) provided energy savings only for day temperatures above 32 °C.

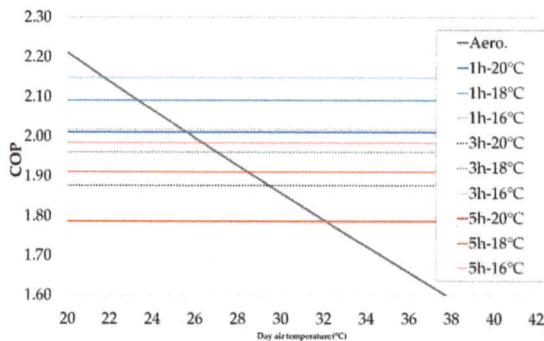

Figure 14. COP of the PCM mode and the aerothermal mode for different use and temperature scenarios.

5. Conclusions

PCM TES systems can provide energy savings when used as a heat sink for a chiller, and there is worth in analyzing the parameters that affect their performance. In the present work, an experimental PCM TES system used as a heat sink for a chiller has been modeled and analyzed under different temperature and use conditions for a better understanding of the behavior of its energy performance.

First, a model of the chiller-PCM components was developed and validated using data from an experimental setup adapted for this purpose. The model showed good accuracy with a maximum deviation between the modeled and the experimental temperatures of 1.2 °C, and it can be used in systems of similar purpose.

Then, this model was used to carry out simulations of the different operating modes of the system providing consumption and COP curves and analyzing the effect of the ambient temperature and the liquid fraction of the PCM. Moreover, the energy savings provided by the system was evaluated for different temperature profiles.

Energy savings between 5% and 15% were observed in the simulations. Furthermore, the system showed the best energy savings profile for a value of T_m around 2 °C below the melting point of the PCM. Simulation results also showed that a minimum difference of 12 °C was necessary between day and night temperature to obtain energy savings with the components used in this study. Although the system showed moderate savings, there is a potential improvement of these results if a higher thermal conductivity between the water and the PCM is achieved. This increase in thermal conductivity, which is the focus of recent studies [25–27], would reduce the water outlet temperature during the discharge mode shown in Figure 7a, and this would increase the COP of the chiller.

Finally, the COP of the aerothermal mode was compared with that of the PCM mode for different use and temperature conditions obtaining, for each case, the switch day ambient temperature over which PCM mode provided energy savings. Considering the model assumptions, for the most profitable conditions, the chiller-PCM system provided savings for day temperatures over 22 °C, whereas under worse conditions, switch temperatures of up to 32 °C were obtained.

Author Contributions: Conceptualization, A.R.F. and J.N.E.; methodology, A.R.F., Á.B.C., and L.D.; validation, A.R.F., J.N.E., A.M.B., and L.D.; formal analysis, A.R.F., J.N.E., A.M.B., and L.D.; investigation, A.R.F., J.N.E., A.M.B., Á.B.C., L.D., F.S., A.M., and C.A.; resources, A.R.F., J.N.E., Á.B.C., and F.S.; data curation, A.R.F. and A.M.B.; writing—original draft preparation, A.R.F., J.N.E., and A.M.B.; writing—review and editing, A.M. and C.A.; visualization, A.M. and C.A.; supervision, J.N.E. and F.S.; project administration, A.R.F., J.N.E., and F.S.; funding acquisition, A.R.F., J.N.E., and F.S.

Funding: A.M.B. acknowledges the funding received from the Spanish State Research Agency through the "Juan de la Cierva - Formación 2016" postdoctoral grant (FJCI-2016-28324).

Acknowledgments: The authors would like to thank the ISTENER research group from Universitat Jaume I and the TEC-ENER research group from Universidad Cardenal Herrera-CEU for providing the equipment required to perform this study.

Conflicts of Interest: The authors declare no conflict of interest.

Nomenclature

A	Daily temperature amplitude (°C)
CR	Consumption ratio (-)
C_p	Specific heat (kJ·kg^{-1}·K^{-1})
d_h	Hydraulic diameter (m)
f	Liquid fraction (-)
h	Film coefficient (W·m^{-2}·K^{-1})
k	Thermal conductivity (W·m^{-1}·K^{-1})
L	Latent heat (kJ·kg^{-1})
m	Mass (kg)
Nu	Nusselt number (-)
P	Power (W)
Pr	Prandtl number (-)
Q	Heat (J)
$\dot{Q}$	Heat transfer rate (W)
R	Thermal resistance (K·W^{-1})
Ra	Rayleigh number (-)
Re	Reynolds number (-)
S	Surface (m^2)
T	Temperature (°C)
T_m	Daily mean temperature (°C)
T_{melt}	Melting temperature (°C)
$\dot{V}$	Volumetric flow rate (m^3·s^{-1})
W	Work (J)

Greek symbols

δ	Dirac delta function
ρ	Density (kg·m^{-3})

Subscripts

aero	Aerothermal mode
cnd	Condenser
ch	Charge mode
comp	Compressor
cont	Container
dch	Discharge mode
ev	Evaporator

hz	Homogenization zone
in	Inlet
liq	Liquid
out	Outlet
p	Pump
sc	PCM section
sol	Solid
w	Water

Abbreviatures

COP	Coefficient of Performance
HVAC	Heating, Ventilation and Air-Conditioning
PCM	Phase Change Material
PID	Proportional Integral Derivative
TES	Thermal Energy Storage
VCS	Vapor Compression System

References

1. UNFCCC. Kyoto Protocol to the United Nations Framework Convention on Climate Change. In Proceedings of the Third Session of the Conference of the Parties (COP3), Kyoto, Japan, 11 December 1997.
2. UNFCCC. *Adoption of the Paris Agreement*; UNFCCC: Bonn, Germany, 2015.
3. Couloumb, D.; Dupont, J.L.; Pichard, A. *29th Informatory Note on Refrigeration Technologies. The Role of Refrigeration in the Global Economy*; International Institute of Refrigeration: Paris, France, 2015.
4. Couloumb, D.; Dupont, J.L.; Morlet, V. *35th Informatory Note on Refrigeration Technologies. The Impact of the Refrigeration Sector on Climate Change*; International Institute of Refrigeration: Paris, France, 2017.
5. International Energy Agency. *Key World Energy Statistics*; International Energy Agency: Paris, France, 2017; Available online: https://www.iea.org/publications/freepublications/publication/KeyWorld2017.pdf.
6. Seong, Y.B.; Lim, J.H. Energy saving potentials of phase change materials applied to lightweight building envelopes. *Energies* **2013**, *6*, 5219–5230. [CrossRef]
7. Baek, S.; Kim, S. Analysis of thermal performance and energy saving potential by PCM radiant floor heating system based on wet construction method and hot water. *Energies* **2019**, *12*, 828. [CrossRef]
8. Roman, K.K.; O'Brien, T.; Alvey, J.B.; Woo, O.J. Simulating the effects of cool roof and PCM (phase change materials) based roof to mitigate UHI (urban heat island) in prominent US cities. *Energy* **2016**, *96*, 103–117. [CrossRef]
9. Anisur, M.R.; Mahfuz, M.H.; Kibria, M.A.; Saidur, R.; Metselaar, I.H.S.C.; Mahlia, T.M.I. Curbing global warming with phase change materials for energy storage. *Renew. Sustain. Energy Rev.* **2013**, *18*, 23–30. [CrossRef]
10. Canbazoglu, S.; Sahinaslan, A.; Ekmekyapar, A.; Aksoya, Y.; Akarsu, F. Enhancement of solar thermal energy storage performance using sodium thiosulfate pentahydrate of a conventional solar water-heating system. *Energy Build.* **2005**, *37*, 235–242. [CrossRef]
11. Charvát, P.; Klimeš, L.; Zálešák, M. Utilization of an air-PCM heat exchanger in passive cooling of buildings: A simulation study on the energy saving potential in different European climates. *Energies* **2019**, *12*, 1133. [CrossRef]
12. Chen, X.; Zhang, Q.; Zhai, Z.J.; Ma, X. Potential of ventilation systems with thermal energy storage using PCMs applied to air conditioned buildings. *Renew. Energy* **2019**, 39–53. [CrossRef]
13. García, J.P.; Miguez, C.; Monedero, C.; Rico, I. *Síntesis del Estudio Parques de Bombas de Calor de España*; IDEA: Madrid, Spain, 2014.
14. Sun, Y.; Wang, S.; Xiao, F.; Gao, D. Peak load shifting control using different cold thermal energy storage facilities in commercial buildings: A review. *Energy Convers. Manag.* **2013**, *71*, 101–114. [CrossRef]
15. Ruddell, B.L.; Salamanca, F.; Mahalov, A. Reducing a semiarid city's peak electrical demand using distributed cold thermal energy storage. *Appl. Energy* **2014**, *134*, 35–44. [CrossRef]
16. Bruno, F.; Tay, N.H.S.; Belusko, M. Minimising energy usage for domestic cooling with off-peak PCM storage. *Energy Build.* **2014**, *76*, 347–353. [CrossRef]

17. Roth, K.W.; Zogg, R.; Brodrick, J. Emerging technologies: Cool thermal energy storage. *ASHRAE J.* **2006**, *48*, 94–96.

18. Zhao, D.; Tan, G. Experimental evaluation of a prototype thermoelectric system integrated with PCM (phase change material) for space cooling. *Energy* **2014**, *68*, 658–666. [CrossRef]

19. Zhao, D.; Tan, G. Numerical analysis of a shell-and-tube latent heat storage unit with fins for air-conditioning application. *Appl. Energy* **2015**, *138*, 381–392. [CrossRef]

20. PCMProducts Ltd. Available online: http://www.pcmproducts.net/Encapsulated_PCMs.htm (accessed on 9 September 2019).

21. Barreneche, C.; Solé, A.; Miró, L.; Martorell, I.; Fernández, A.I.; Cabeza, L.F. Study on differential scanning calorimetry analysis with two operation modes and organic and inorganic phase change material (PCM). *Thermochim. Acta* **2013**, *553*, 23–26. [CrossRef]

22. Dittus, F.; Boelter, L. *Heat Transfer in Automobile Radiators of the Tubular Type*; University of California Publications in Engineering; University of California Press: Berkeley, CA, USA, 1930; Volume 2, pp. 443–461.

23. Chiu, J.N.W. Latent Heat Thermal Energy Storage for Indoor Comfort Control. Ph.D. Thesis, KTH School of Industrial Engineering and Management, Stockholm, Sweden, 2013.

24. Kuznik, F.; Virgone, J.; Johannes, K. Development and validation of a new TRNSYS type for the simulation of external building walls containing PCM. *Energy Build.* **2010**, *42*, 1004–1009. [CrossRef]

25. Mat, S.; Al-Abidi, A.A.; Sopian, K.; Sulaiman, M.Y.; Mohammad, A.T. Enhance heat transfer for PCM melting in triplex tube with internal-external fins. *Energy Convers. Manag.* **2013**, *74*, 223–236. [CrossRef]

26. Righetti, G.; Lazzarin, R.; Noro, M.; Mancin, S. Phase Change Materials embedded in porous matrices for hybrid thermal energy storages: experimental results and modelling. *Int. J. Refrig.* **2019**. [CrossRef]

27. Lazzarin, R.; Noro, M.; Righetti, G.; Mancin, S. Application of hybrid PCM thermal energy storages with and without al foams in solar heating/cooling and ground source absorption heat pump plant: An energy and economic analysis. *Appl. Sci.* **2019**, *9*, 1007. [CrossRef]

energies

MDPI

Article

Enhancement of a R-410A Reclamation Process Using Various Heat-Pump-Assisted Distillation Configurations

Nguyen Van Duc Long [1,†], Thi Hiep Han [1,†], Dong Young Lee [1], Sun Yong Park [2], Byeng Bong Hwang [2] and Moonyong Lee [1,*]

[1]   School of Chemical Engineering, Yeungnam University, Gyeongsan 712-749, Korea;
      allenthelong@yu.ac.kr (N.V.D.L.); hanthihiep@yu.ac.kr (T.H.H.); ldy5525@ynu.ac.kr (D.Y.L.)
[2]   OunR2tech Co., Ltd, Pohang 37553, Korea; psy6313@naver.com (S.Y.P.); ounr2tech@naver.com (B.B.H.)
*   Correspondence: mynlee@yu.ac.kr; Tel.: +82-53-810-2512
†   These two authors contributed equally to this work.

Received: 15 September 2019; Accepted: 2 October 2019; Published: 4 October 2019

Abstract: Distillation for R-410A reclamation from a waste refrigerant is an energy-intensive process. Thus, various heat pump configurations were proposed to enhance the energy efficiency of existing conventional distillation columns for separating R-410A and R-22. One new heat pump configuration combining a vapor compression (VC) heat pump with cold water and hot water cycles was suggested for easy operation and control. Both advantages and disadvantages of each heat pump configuration were also evaluated. The results showed that the mechanical vapor recompression heat pump with top vapor superheating saved up to 29.5%, 100.0%, and 10.5% of the energy required in the condenser duty, reboiler duty, and operating cost, respectively, compared to a classical heat pump system, and 85.2%, 100.0%, and 60.8%, respectively, compared to the existing conventional column. In addition, this work demonstrated that the operating pressure of a VC heat pump could be lower than that of the existing distillation column, allowing for an increase in capacity of up to 20%. In addition, replacing the throttle valve with a hydraulic turbine showed isentropic expansion can decrease the operating cost by up to 20.9% as compared to the new heat pump configuration without a hydraulic turbine. Furthermore, the reduction in carbon dioxide emission was investigated to assess the environmental impact of all proposed sequences.

Keywords: distillation; heat pump; hydraulic turbine; refrigerant reclamation; R-410A; superheating

1. Introduction

Traditional refrigerants including hydrochlorofluorocarbons (HCFCs) (e.g., chlorodifluoromethane CHClF2 (R-22)) and chlorofluorocarbons (CFCs) can cause global warming and ozone depletion owing to their poor stability in the atmosphere. Therefore, they are being replaced by other environmentally friendly products, such as hydrofluorocarbons (HFCs) (e.g., R-410A) [1–3]. R-410A, which is an azeotropic mixture of difluoromethane (R-32) and pentafluoroethane (R-125), has been developed as a long-term replacement for R-22 [4]. Although they provide an effective alternative to HCFCs and CFCs, they are still greenhouse gases with long atmospheric lifetimes.

Refrigerant reclamation is defined as the processing of used refrigerant gases so that they can meet specifications for re-use. Reclamation services were designed with the objective of reducing the environmental impact of used refrigerants, allowing for the recycling of existing refrigerants, thereby avoiding the need to manufacture new refrigerant molecules [5]. One of the main challenges in refrigerant reclamation is to separate and purify refrigerant mixtures because they usually form azeotropic mixtures or close-boiling mixtures [6]. Therefore, in view of the current restrictions associated with refrigerants, developing new, effective, and efficient methods to separate and purify individual components from an azeotropic mixture is crucial.

A few techniques have recently been developed to separate R-410A and R-22. Wanigarathna et al. proposed an adsorption process for the separation of mixtures of R-32, R-22, and R-125 using 4A molecular sieve zeolite [7]. The expensive equipment (material) used in adsorption due to the need for a considerable amount of stationary phase material is also significant. The distillation process, which is the most common process employed in the chemical and petrochemical industries, is used for this separation in OunR2tech company in Korea. However, distillation for separating R-410A and R-22 is an energy-intensive process. There is a need to enhance the energy efficiency of this process and to decrease direct emissions of greenhouse gases, which have a considerable impact on climate change.

Distillation is a well-developed, thermal-driven separation process and is extensively used in the chemical industry; however, it consumes a massive amount of fossil fuel as the heat source. There are several ways to improve the performance of distillation as this will lead to a reduction in the consumption of fossil fuel, thereby leading to a significant carbon dioxide (CO_2) emissions reduction. One of the most attractive solutions is using cleaner electricity generated from sustainable energy sources, such as wind power and solar cells. Heat pump (HP) technology, which allows for the use of the waste heat released at the condenser for supplying energy to the reboiler, is a cost-effective approach to reducing the energy requirement of the distillation column [8–11]. This technology can be beneficial in the case of the small temperature difference between the overhead and bottom of the column and the high heat load. To reduce the energy requirement of the distillation columns, several HP concepts have been developed. Among them, the mechanical vapor recompression (MVR) heat pump and vapor compression (VC) heat pump are the most popular configurations. As the heat is exchanged only once in an MVR, along with a smaller condenser and the lower temperature lift, is advantageous over a VC, this results in a higher thermodynamic efficiency [12].

Considering all the above-mentioned issues, in this study, various HP configurations were proposed for enhancing the energy efficiency of the existing conventional distillation for R-410A reclamation from a waste refrigerant. Top vapor superheating was proposed for improving the performance of the HP configuration, as well as for protecting the compressor from liquid leakage. One new HP configuration combining the VC with cold water and hot water cycles was suggested for easy operation and control. Furthermore, hydraulic turbines were evaluated to enhance the performance of the new HP configuration. The reduction in CO_2 emissions was also determined as part of the evaluation of all proposed sequences.

2. Proposed Configurations for Improving Reclamation Process

2.1. Mechanical Vapor Recompression

MVR (Figure 1a) can be applied in many industrial processes, such as crystallization, evaporation, stripping, drying, and distillation [13–15]. In such a system, the overhead vapor is fed directly to the compressor and compressed to a level that can transfer the heat to the reboiler, leading to a significant reduction in the energy consumption of both the condenser and reboiler [16–21]. An HP is desired if the coefficient of performance (COP) or the ratio of the amount of heat delivered to the reboiler (Q) and the external work (W) is more than 10, while further evaluation is necessary if this value is between 5 and 10 and should not be considered when the Q/W ratio is lower than 5 [21,22].

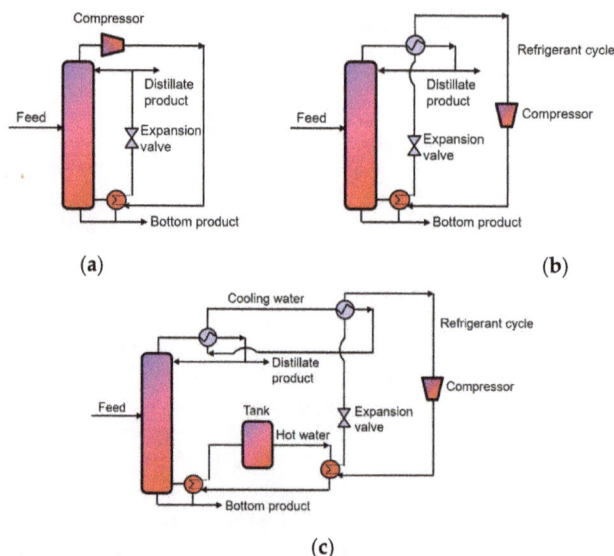

Figure 1. Schematic of the heat pump systems: (**a**) mechanical vapor recompression heat pump, (**b**) vapor recompression heat pump, and (**c**) new heat pump system.

2.2. Vapor Compression Heat Pump

VCs are the most common type of HPs with a limited current maximum temperature of 120 °C [15]. In VC (Figure 1b), a specific working fluid is used as a heat transfer medium. The main characteristic of VC is the phase change of this fluid. In particular, the working fluid obtains the waste heat from the low-temperature overhead vapor stream to convert the working fluid to vapor before being compressed by a compressor to increase its pressure and temperature. The compressed fluid then supplies energy to the bottom products to boil the column up and be changed to a liquid phase [13]. In the next stage, the fluid enters an expansion valve to reduce its pressure and temperature before continuing another cycle. To achieve a good performance, the most important parameter is the choice of suitable working fluid.

2.3. New Configuration

In VC heat pumps, a buffer tank containing a certain amount of refrigerant is needed. During low-load conditions, the extra volume of refrigerant would absorb any of the extra heat generated by the HP. However, the refrigerant is quite expensive. Thus, in the cases where the difference between the top and bottom temperatures is less, a new configuration containing a cheap water buffer tank can be considered. Figure 1c shows the process flow diagram of the proposed new configuration. This can be considered as the modified VC configuration with one hot water cycle and one cold water cycle.

In this new configuration, the operating conditions of the HP cycle are modified such that the temperature of the refrigerant is lower than that of cooling water. After obtaining the waste heat from the overhead vapor, the cooling water transfers the heat to the refrigerant in a heat exchanger. This heat totally generates refrigerant vapor and superheats it. Afterward, the superheated refrigerant is introduced into a compressor to convert it to a higher temperature and pressure. The high-temperature refrigerant is then cooled down to its dew point, condensed, and cooled below its boiling point by releasing thermal energy to provide the heat to the hot water used to boil the column up. Before completing the cycle, this refrigerant is passed through a valve to reduce its pressure, temperature, and bubble point temperature.

Creating temperature differences between the cooling water and overhead vapor, the cooling water and the refrigerant, the refrigerant and the hot water, and the hot water and the bottom product is necessary. Therefore, supplying a minimum temperature approach in the heat exchangers required a larger compression ratio value of the compressor. Thus, the energy and economic performance of our proposed configuration is usually lower than that of VC and MVR. Nevertheless, as this configuration only has a water buffer tank, it is cheaper than the system containing the expensive refrigerant buffer tank. Furthermore, the performances of this system can be higher if cheaper electricity is produced in some companies or countries and cleaner electricity obtained from sustainable energy sources are used to drive the compressor of the HP system. Besides, implementation, operation, and control become simple.

3. Results and Discussion

3.1. Existing Conventional Distillation Column

The existing conventional distillation column operated at 27.7 bar possesses 49 theoretical stages. The flowchart of the existing distillation column for the R-410A reclamation from a waste refrigerant is presented in Figure 2. The light product R-410A (purity: 99.5%) was obtained from the top of the column. Aspen HYSYS V10 (Aspen Technology, Inc., Bedford, MA, United States) was employed to simulate all configurations in this study. The REFPROP property method provided thermodynamic and transport properties of industrially important fluids with an emphasis on refrigerants [23]. Thus, REFPROP was selected as the property package for all simulations in this study. The reboiler duty of the existing distillation column was 48.15 kW (shown in Table 1).

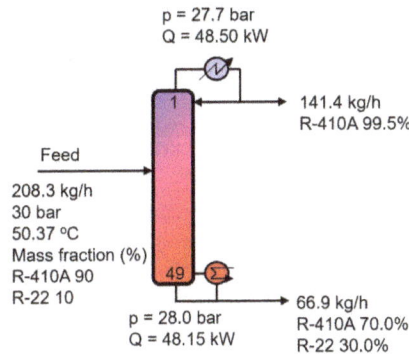

Figure 2. Simplified flowchart illustrating the existing conventional column for R-410A reclamation.

Table 1. Existing conventional distillation columns' hydraulics, energy performance, and product specifications.

Number of trays	49
Tray type	Packing
HETP (m)	0.43
Column diameter (m)	0.25
Max flooding (%)	68.3
Energy requirement of condenser (kW)	48.50
Energy requirement of reboiler (kW)	48.15
R-410A purity (mass%)	99.5

3.2. MVR Heat Pump

In the existing column, the COP value of 8.5 revealed that the heat released in the condenser can be utilized to supply the reboiler through an MVR system. The flowchart for the MVR-assisted distillation column is shown in Figure 3. In this configuration, an extra reboiler was utilized because the amount of energy transferred was insufficient. Compared to the existing conventional distillation column, the use of an MVR heat pump could improve the reboiler duty and operating costs by 94.3% and 56.2%, respectively, as shown in Table 2. Note that the costs of low-pressure steam, cooling water, and electricity were 6.08 \$/GJ, 0.35 \$/GJ, and 16.80 \$/GJ, respectively, which were used for the calculation of the total operating cost [24].

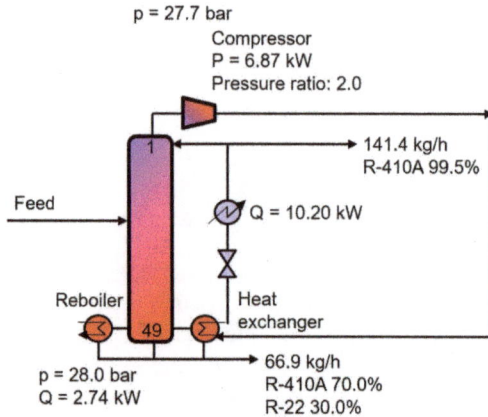

Figure 3. Simplified flowchart illustrating the MVR heat-pump-assisted conventional column.

In some cases, there is partial condensation of saturated vapor in the compressor, causing a liquid break. To avoid this problem, the top vapor was modified to pass through the compressor after it was superheated by the remaining heat in the exchanger outlet stream in a superheater (shown in Figure 4) [14,25]. In this configuration, both latent heat and sensible heat were used to supply the heat to the reboiler and superheat the top vapor stream. Because of preheating of the top product, a lower remaining reboiler duty and condenser duty were required. Specifically, the remaining reboiler duty was decreased from 2.84 kW to 0 kW, i.e., this column was driven by self-heat recuperation. In addition, the remaining condenser duty was reduced from 10.20 kW to 7.19 kW, i.e., more savings were achieved by adding a heat exchange stage to heat the top vapor stream of the column before passing through the compressor. Furthermore, the pressure ratio was reduced slightly from 2.0 to 1.9.

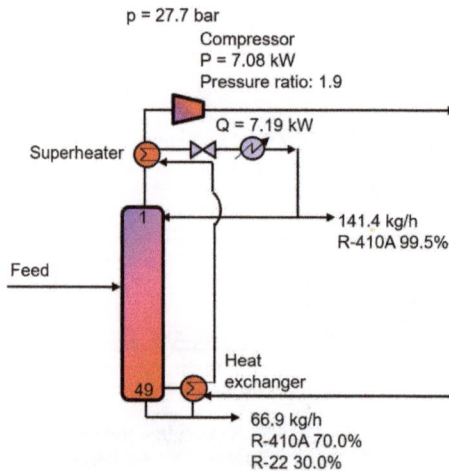

Figure 4. Simplified flowchart illustrating the MVR heat-pump-assisted conventional column with top vapor superheating.

This modification can induce savings of 29.5% and 100.0% in terms of condenser duty and reboiler duty, respectively, as compared to the MVR-assisted distillation without superheating. Moreover, using this sequence can save 60.8% in terms of the operating cost compared to the existing distillation column. These savings can be higher if cleaner electricity obtained from sustainable energy sources, such as wind or solar power, is used to drive the compressor of the HP system.

3.3. VC Heat Pump

The selection of an appropriate working fluid is an important parameter in the design of a VC heat pump configuration. In this work, to select the working fluid, the following criteria were considered [25]:

1. The bubble point of the working fluid at 1.01 bar must be less than the temperature of the top vapor by at least 10 °C.
2. The dew point of the working fluid at higher pressure should be higher than the reboiler temperature by at least 10 °C.

Under the operating conditions of the distillation column, R-22 is a good working fluid. Nevertheless, R-407C, which is a blend of R-32, R-125, and 1,1,1,2-tetrafluoroethane (R-134A), is a similar refrigerant to R-22 in terms of working pressure, capacity, and energy efficiency for above-zero and mid-range evaporation temperatures [26]. R-407C has emerged as the popular choice for supermarket refrigeration equipment because of its good performance match to R-22 and its lower global warming potential than other alternatives [27]. Thus, R-407C was selected as the fluid for the refrigeration cycle.

The adiabatic efficiency of the compressor was assumed to be 75% and the minimum temperature approach in the heat exchanger was assumed to be 10 °C. Note that the superheating level was 3 °C and the subcooling level was 2 °C. The parameters affecting the overall utility energy consumption were the pressure of the compressor inlet and outlet streams and the flow rate of the working fluid [28]. They were adjusted to satisfy the minimum temperature approach of 10 °C in each heat exchanger of the system and superheating level of 3 °C. Because the heat capacity of the HP system exceeded the reboiler duty, one dry cooler was installed to release the extra heat and achieve a subcooling level of 2 °C.

The flowchart of the VC-assisted distillation column is shown in Figure 5. Interestingly, because the VC heat pump configuration did not have a condenser using cooling water, the column could be operated at a lower operating pressure. As indicated in Table 2, the VC-assisted distillation column can save more reboiler duty and condenser duty than an MVR-assisted distillation configuration. However, heat was exchanged twice, and the temperature increase in the VC heat pump was 10 °C higher than that of the MVR system. This induced a higher compression duty, causing lower operating cost savings (54.5%) compared to 56.2% when using MVR.

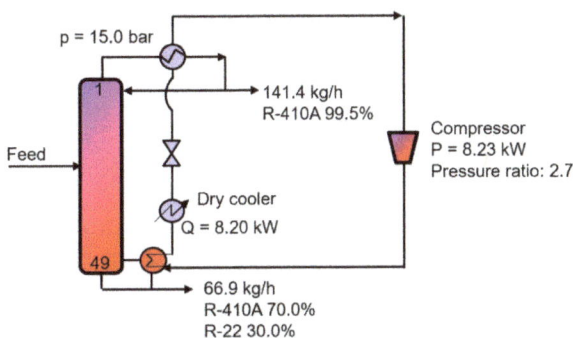

Figure 5. Simplified flowchart illustrating the VC heat-pump-assisted conventional column.

Note that operating at a lower pressure reduced the energy required to drive the column of the VC heat-pump-assisted distillation, which facilitated an increase in capacity. In particular, the capacity of this configuration could be increased by up to 20% compared to the existing conventional distillation and MVR heat pump system. In addition, the total annual CO_2 emissions calculated using Gadalla's modular method [29] decreased by up to 50.7% compared to the existing conventional distillation column.

3.4. New Heat Pump Configuration

Another configuration including a refrigerant cycle, a hot water cycle, and a cold water cycle was also considered to improve the existing conventional distillation column (shown in Figure 6). Note that the cooling water temperature was 7 °C, whereas hot water temperature was 45 °C. The thermodynamic cycle of the new configuration is shown in Figure 7. The superheated fluid (1: T = 1.1 °C, P = 4.3 bar) entered the compressor to increase its pressure. Then, the compressed working fluid (2: T = 79.5 °C, P = 20.3 bar) was first cooled down to its dew point, condensed, and cooled below its boiling point (3: T = 44.2 °C, P = 20.3 bar). Next, the pressure and temperature of this working fluid were decreased using a Joule–Thompson (JT) throttling valve (4: T = −6.1 °C, P = 4.3 bar). Finally, the vapor–liquid mixture was totally evaporated and superheated in the heat exchanger with cooling water. Because compressor suction pressure P_{in} (4.3 bar) was higher than the atmospheric pressure, it could be reduced easily.

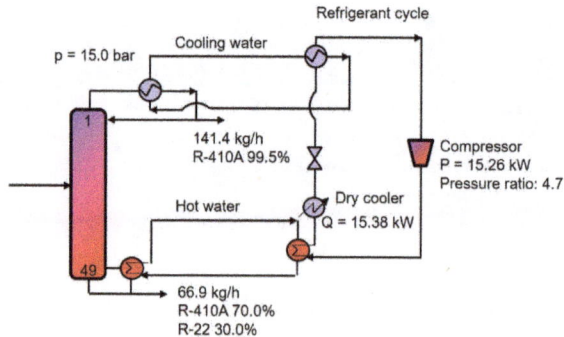

Figure 6. Simplified flowchart illustrating the new heat pump assisted conventional column.

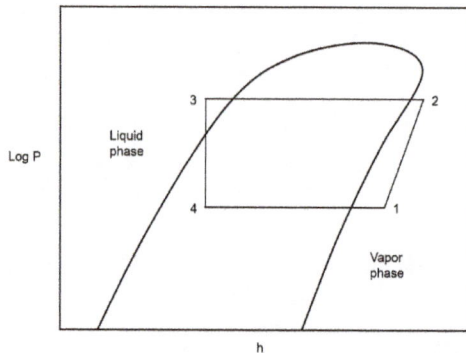

Figure 7. Thermodynamic cycle for the vapor compression of the new heat-pump-assisted conventional column.

Compared to VC, the high temperature difference between the low pressure and high pressure working fluids created a need for large compression ratios of the compressor. As a result, despite the high savings in terms of condenser duty (68.3%) and reboiler duty (100.0%), the operating cost savings were only 15.6% as compared to the existing conventional distillation column.

A JT throttling valve is the most widely used expansion device in refrigerant cycles because of its many practical advantages, such as simplicity, and low investment and maintenance costs [30]. However, from a thermodynamic point of view, it has an inherent limitation, which is a low expansion efficiency because an expansion process through the JT valve essentially consists of an isenthalpic process. On the other hand, in the case of a hydraulic turbine expansion (also known as a liquid expander)-based refrigeration cycle, the liquid portion of the working fluid will be higher than that employing the JT valve. Thus, the isentropic expansion will result in a higher cooling capacity per unit mass of mixed refrigerant with lower shaft work [30–34]. Recent technological advancements in cryogenic liquid expansion turbines have enabled the replacement of the JT expansion valve with a cryogenic power recovery turbine [30,31,35]. This was used to improve liquefied natural gas operations [30,36,37].

Thus, the hydraulic turbine was investigated to replace the JT valves and consequently improve the energy efficiency of the HP cycle (shown in Figure 8). As a result, the outlet of the JT valve had the same molar enthalpy as the inlet stream, whereas the outlet of hydraulic turbine had a higher molar enthalpy (7.790×10^5 kJ/kg·mol) than that of the inlet stream (7.785×10^5 kJ/kg·mol). Furthermore, the temperature and vapor portions of the working fluid in the outlet stream were lower than those using the JT valve. This resulted in a higher minimum temperature approach between the cold water and the working fluid, consequently increasing the cooling capacity per unit mass of working fluid and decreasing the compressor duty. In particular, the compressor duty was reduced from 15.26 kW to 13.72 kW. Additionally, 1.65 kW could be generated by using the hydraulic turbine, which could be used to drive the compressor. Thus, the hydraulic turbine not only increased the efficiency of the HP system by performing the isentropic throttling function ideally, but also recovered the work of expansion. Consequently, the operating cost savings could be up to 20.9% and 33.2% as compared to the new HP configuration without a hydraulic turbine and an existing distillation column, respectively.

Table 2 summarizes the results concerning the operating cost and CO_2 emission performance comparisons between all heat pump arrangements studied in this work and the existing conventional column. The performances of these configurations can be higher if cheaper electricity is produced in the companies or countries in question and cleaner electricity is obtained from sustainable energy sources, such as wind or solar power, are used to drive the compressor of the HP system. To select the suitable configuration, the decision-maker should consider not only the energy performance and environmental impact, but also the operability of the system. The main purpose of this paper was to show the feasibility of the enhancement of the R-410A reclamation process using various heat-pump-assisted distillation configurations. We would like to save operability analysis and proposed control structure of these configurations for the next paper.

Table 2. Comparison of different structural alternatives.

Structural Alternative	Existing Conventional Column	MVR Heat Pump	MVR Heat Pump with Superheating	VC Heat Pump	New Heat Pump	New Heat Pump with Hydraulic Turbine
Energy requirement saving in condenser (%)	-	79.0	85.2	83.1	68.3	74.9
Energy requirement saving in reboiler (%)	-	94.3	100.0	100.0	100.0	100.0
Annual operating cost saving (%)	-	56.2	60.8	54.5	15.6	33.2
CO_2 emission saving (%)	-	53.2	57.6	50.7	8.6	27.7

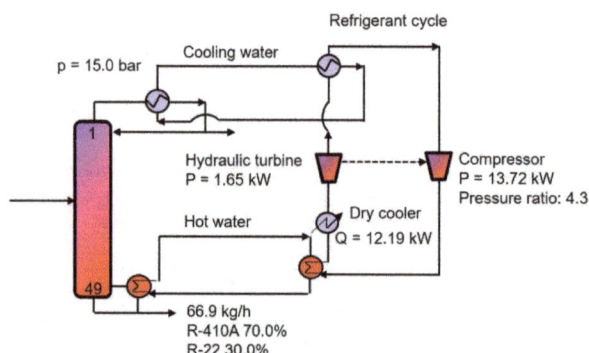

Figure 8. Simplified flowchart illustrating the new heat-pump-assisted conventional column with a hydraulic turbine.

4. Conclusions

The purpose of this study was to propose different heat-pump-assisted distillation configurations for improving the energy efficiency of the R-410A reclamation process from a waste refrigerant. Top vapor superheating was proved to be a promising option for improving the performance of HP configurations, as well as protecting the compressor from the liquid break. This modification could induce savings of 29.5% and 100.0% in terms of condenser duty and reboiler duty, respectively, compared to MVR-assisted distillation without superheating. As a result, using this sequence could save 60.8% in terms of operating cost compared to the existing distillation column. These savings can be higher if cheaper electricity produced in some companies or countries and cleaner electricity obtained from sustainable energy sources, such as wind or solar power, are used to drive the compressor of the HP system. This study also demonstrated that the VC heat pump could be operated at a lower operating pressure than the existing distillation column and this could induce an increase in capacity of up to 20%. Furthermore, this work proposed an effective HP configuration that was easy to operate and control. The hydraulic turbine not only increased the efficiency of the HP system by performing the isentropic throttling function ideally, but could also recover the work of expansion. The results also indicated that the CO_2 emission could be reduced significantly when enhancing the existing conventional distillation column using HP configurations.

Author Contributions: N.V.D.L. and T.H.H. conducted the main job of modeling and designing all configurations. D.Y.L. aided in designing the enhanced configuration. S.Y.P. and B.B.H. aided in the design of the new configuration. M.L. conceived the core concepts for the research and provided academic advice. All authors collaborated for the preparation, revisions, and general editing of this manuscript.

Funding: This work was supported by the Basic Science Research Program through the National Research Foundation of Korea (NRF) funded by the Ministry of Education (2018R1A2B6001566), and by Priority Research Centers Program through the National Research Foundation of Korea (NRF) funded by the Ministry of Education (2014R1A6A1031189), and the R&D Center for Reduction of Non-CO_2 Greenhouse Gases (201700240008) funded by the Ministry of Environment as a "Global Top Environment R&D Program."

Conflicts of Interest: The authors declare no conflict of interest.

References

1. Summary of Refrigerant Reclamation Trends. Available online: https://www.epa.gov/section608/summary-refrigerant-reclamation (accessed on 11 August 2019).
2. Available online: https://www.agas.co.uk/products-services/recovery-reclamation-disposal (accessed on 11 August 2019).
3. Refrigerant reclaim solutions. Available online: http://www.linde-gas.com/en/images/Refrigerant%20Reclaim%20Solutions%20brochure_tcm17-108595.pdf (accessed on 11 August 2019).

4. Refrigerant 410A. Available online: https://www.rses.org/assets/r410a/R-410A.PDF (accessed on 11 August 2019).

5. Available online: https://www.boconline.co.uk/en/products-and-supply/refrigerant-gases/reclaim-recovery-and-waste-management/index.html (accessed on 11 August 2019).

6. Gregorio, T. Apparatus and Process for the Separation and Purification of Ideal and Non-Ideal Refrigerant Mixtures. US8075742, 13 December 2011.

7. Wanigarathna, D.J.A.; Gao, L.; Takanami, T.; Zhang, Q.; Liu, B. Adsorption separation of R-22, R-32 and R-125 fluorocarbons using 4A molecular sieve zeolite. *Chem. Select* **2016**, *1*, 3718–3722.

8. Chew, J.M.; Reddy, C.C.S.; Rangaiah, G.P. Improving energy efficiency of dividing-wall columns using heat pumps, organic rankine cycle and kalina cycle. *Chem. Eng. Process* **2014**, *76*, 45–59. [CrossRef]

9. Long, N.V.D.; Minh, L.Q.; Nhien, L.C.; Lee, M. A novel self-heat recuperative dividing wall column to maximize energy efficiency and column throughput in retrofitting and debottlenecking of a side stream column. *Appl. Energy* **2015**, *159*, 28–38. [CrossRef]

10. Minh, L.Q.; Long, N.V.D.; Duong, P.L.T.; Jung, Y.; Bahadori, A.; Lee, M. Design of an extractive distillation column for the environmentally benign separation of zirconium and hafnium tetrachloride for nuclear power reactor applications. *Energies* **2015**, *8*, 10354–10369. [CrossRef]

11. Lee, J.; Son, Y.; Lee, K.S.; Won, W. Economic analysis and environmental impact assessment of heat pump-assisted distillation in a gas fractionation unit. *Energies* **2019**, *12*, 852. [CrossRef]

12. Bruinsma, D.; Spoelstra, S. Heat pumps in distillation. In Proceedings of the Distillation & Absorption Conference, Eindhoven, The Netherlands, 12–15 September 2010.

13. Kazemi, A.; Mehrabani-Zeinabad, A.; Beheshti, M. Recently developed heat pump assisted distillation configurations: A comparative study. *Appl. Energy* **2018**, *211*, 1261–1281. [CrossRef]

14. Long, N.V.D.; Lee, M. *Advances in Distillation Retrofit*, 1st ed.; Springer: New York, NY, USA, 2017.

15. Kiss, A.A.; Ferreira, C.A.I. *Heat Pumps in Chemical Process Industry*, 1st ed.; CRC Press: Boca Raton, FL, USA, 2017.

16. Schinitzer, H.; Moser, F. *Heat Pumps in Industry*; Elsevier: Amsterdam, The Netherlands, 1985.

17. Ranade, S.; Chao, Y. Industrial heat pumps: Where and when? *Hydrocarb. Process* **1990**, 71–73.

18. Mizsey, P.; Fonyo, Z. Energy integrated distillation system design enhanced by heat pumping. *Distill. Absorpt.* **1992**, 1369–1376.

19. Annakou, O.; Mizsey, P. Rigorous investigation of heat pump assisted distillation. *Heat Recov. Syst. CHP* **1995**, *15*, 241–247. [CrossRef]

20. Long, N.V.D.; Lee, M. Review of retrofitting distillation columns using thermally coupled distillation sequences and dividing wall columns to improve energy efficiency. *J. Chem. Eng. Jpn.* **2014**, *47*, 87–108. [CrossRef]

21. Long, N.V.D.; Lee, M. Novel acid gas removal process based on self-heat recuperation technology. *Int. J. Greenh. Gas Control* **2017**, *64*, 34–42. [CrossRef]

22. Pleşu, V.; Ruiz, A.E.B.; Bonet, J.; Llorens, J. Simple equation for suitability of heat pump use in distillation. *Comput. Aided Chem. Eng.* **2014**, *33*, 1327–1332.

23. Aspen Technology. *Aspen Physical Property System*; Aspen Technology: Bedford, MA, USA, 2013.

24. Turton, R.; Bailie, R.C.; Whiting, W.B.; Shaeiwitz, J.A.; Bhattacharyya, D. *Analysis, Synthesis and Design of Chemical Processes*, 4th ed.; Pearson Education: Upper Saddle River, NJ, USA, 2012.

25. Modla, G.; Lang, P. Heat pump systems with mechanical compression for batch distillation. *Energy* **2013**, *62*, 403–417. [CrossRef]

26. R-407C. Available online: https://www.gas-servei.com/productos/refrigerantes/refrigerantes-hfc/gasficha/r-407c/ (accessed on 11 August 2019).

27. R407 A. Available online: http://www.refrigerants.com/ (accessed on 11 August 2019).

28. Kazemi, A.; Faizi, V.; Mehrabani-Zeinabad, A.; Hosseini, M. Evaluation of performance of heat pump assisted distillation of ethanol-water mixture. *Sep. Sci. Technol.* **2017**, *52*, 1387–1396. [CrossRef]

29. Gadalla, M.A.; Olujic, Z.; Jansens, P.J.; Jobson, M.; Smith, R. Reducing CO_2 emissions and energy consumption of heat-integrated distillation systems. *Environ. Sci. Technol.* **2005**, *39*, 6860–6870. [CrossRef] [PubMed]

30. Muhammad, A.Q.; Ali, W.; Long, N.V.D.; Khan, M.S.; Lee, M. Energy efficiency enhancement of a single mixed refrigerant LNG process using a novel hydraulic turbine. *Energy* **2018**, *144*, 968–976.

31. Kanoğlu, M. Cryogenic turbine efficiencies. *Exergy Int. J.* **2001**, *1*, 202–208. [CrossRef]

32. Minh, N.Q.; Hewitt, N.J.; Eames, P.C. Improved vapour compression refrigeration cycles: Literature review and their application to heat pumps. In Proceedings of the International Refrigeration and Air Conditioning Conference, Purdue University, IN, USA, 17–20 July 2006.

33. Wang, H.; Peterson, R.; Harada, K.; Miller, E.; Ingram-Goble, R.; Fisher, L.; Yih, J.; Ward, C. Performance of a combined organic rankine cycle and vapor compression cycle for heat activated cooling. *Energy* **2011**, *36*, 447–458. [CrossRef]

34. Ferrara, G.; Ferrari, L.; Fiaschi, D.; Galoppi, G.; Karellas, S.; Secchi, R.; Tempesti, D. A small power recovery expander for heat pump COP improvement. *Energy Procedia* **2015**, *81*, 1151–1159. [CrossRef]

35. Gordon, J.L. hydraulic turbine efficiency. *Can. J. Civ. Eng.* **2001**, *28*, 238–253. [CrossRef]

36. Johnson, L.L.; Renaudin, G. Liquid turbines improve LNG operations. *Oil Gas J.* **1996**, *94*, 31–32.

37. Kimmel, H.E.; Cathery, S. Thermo-fluid dynamics and design of liquid-vapour two-phase LNG expanders. In Proceedings of the GPA Europe Technical Meeting, Paris, France, 24–26 February 2010.

energies

MDPI

Article

HFO1234ze(e) As an Alternative Refrigerant for Ejector Cooling Technology

Van Vu Nguyen [1,*], Szabolcs Varga [2] and Vaclav Dvorak [1]

[1] Department of Power Engineering Equipment, Faculty of Mechanical Engineering,
    Technical University of Liberec, Studentska 2, 46117 Liberec, Czech Republic
[2] Department of Mechanical Engineering, INEGI/University of Porto, Rua Dr. Roberto Frias,
    4200-465 Porto, Portugal
* Correspondence: nguyen.van.vu@tul.cz

Received: 30 September 2019; Accepted: 22 October 2019; Published: 24 October 2019

Abstract: The paper presented a mathematical assessment of selected refrigerants for the ejector cooling purpose. R1234ze(e) and R1234yf are the well-known refrigerants of hydrofluoroolefins (HFOs), the fourth-generation halocarbon refrigerants. Nature working fluids, R600a and R290, and third-generation refrigerant of halocarbon (hydrofluorocarbon, HFC), R32 and R152a, were selected in the assessment. A detail mathematical model of the ejector, as well as other components of the cycle, was built. The results showed that the coefficient of performance (COP) of R1234ze(e) was significantly higher than R600a at the same operating conditions. R1234yf's performance was compatible with R290, and both were about 5% less than the previous two. The results also indicated that R152a offered the best performance among the selected refrigerants, but due to the high value of global warming potential, it did not fulfill the requirements of the current European refrigerant regulations. On the other hand, R1234ze(e) was the most suitable working fluid for the ejector cooling technology, thanks to its overall performance.

Keywords: r1234yf; r1234ze(e); HFO; ejector refrigeration technology

---

1. Introduction

The most attractive point of the ejector cooling technologies is that they are driven mainly by heat energy sources. Therefore, they can significantly save electrical energy compared to traditional compressor systems. In a study on a solar-driven ejector cooling system, Guo et al. [1] claimed that the system could save up to 80% electric energy compared to traditional compressor refrigeration. Moreover, ejector technology does not require electrical or mechanical shaft energy input, which significantly reduces equipment mass and increases the reliability of the system.

Up till now, researchers have studied the ejector cooling system with all common refrigerants as working fluids, which can be classified into three groups based on chemical compounds: (i) The halocarbons group: chlorofluorocarbons (CFCs), hydrochlorofluorocarbons (HCFCs), hydrofluorocarbons (HFCs), hydrofluoroolefins (HFOs). (ii) Organic compounds: consist of hydrogen and carbon. (iii) Inorganic refrigerants.

(i) The halocarbons refrigerants have been used widely, thanks to their advantages: can be used for cooling to sub-zero temperatures; they can also operate at low pressures. Here is a quick look about common compounds of this group:

Chlorofluorocarbons (CFCs) and hydrochlorofluorocarbons (HCFCs), had been widely used but then were banned entirely due to their high ozone depletion potential (ODP) and global warming potential (GWP) index. On the other hand, hydrofluorocarbons (HFCs) have zero ODP index and lower GWP than the two others. HFCs also offer favorable performances, and thus they are used in almost every refrigeration application nowadays. There are many studies on ejector cooling technologies using

HFCs in literature; Huang et al. [2] was known as one of the pioneers in ejector refrigeration technology, he mainly used refrigerant R141b; other well-known researchers, such as Khalil et al. [3], Yu and Li [4] used R134a. Unfortunately, under the recent European regulation 517/2014, these refrigerants are being phased out and eventually banned [5].

Hydrofluoroolefins (HFOs) are similar to HFCs as they are composed of hydrogen, fluorine, and carbon. The difference is HFOs are unsaturated: they have at least one double bond. R1234ze(e) and R1234yf are two well-known HFOs that are used in several applications in recent years. However, only a few studies on ejector cooling technology (ECT) using HFOs were found in literature and all used the so-called "drop-in replacement" method: Fang et al. [6] replaced R134a by R1234ze(e), and Smierciew et al. [7] dropped-in R1234ze(e) to a system initially designed for R600a.

(ii) Organic refrigerants (hydrocarbons), on the other hand, have zero ODP and negligible GWP. Isobutane (R600a) and propane have been implemented [8,9]. Despite good thermophysical properties and low GWP, the use of these working fluids is limited due to their high flammability.

(iii) Inorganic refrigerants: Three typical working fluids of this group are ammonia, water, and carbon dioxide. Sankarlal et al. [10] presented an experimental study on the ejector refrigeration system (ERS) with ammonia as working fluid because it was believed that this fluid is environmental-friendly, with good thermodynamics and transport properties compared to some halocarbons. In contrast, it requires high working pressure, and the results of the study did not seem to be desirable. Also, it is a highly toxic refrigerant. Carbon dioxide is another inorganic refrigerant, which has zero impact on the environment, it is non-toxic, non-flammable, and most importantly, it has good efficiency. In particular, by using transcritical cycles, carbon dioxide can achieve excellent performance (coefficient of performance, COP = 3 ÷ 6) [11]. Nevertheless, it is challenging for system construction, as well as operation, since the ERS using carbon dioxide operates at high pressures, up to 10 MPa as claimed in work [12].

As mentioned, the current refrigerant regulation 517/2014 has caused the refrigerant substitute on a large scale. Generally speaking, the regulation was designed to mitigate climate change and protect the environment by reducing emissions of fluorinated greenhouse gases: all F-gases with GWP of 150 or more are not allowed to be used in new installations. The regulation mainly affects the HFCs group, because CFCs and HCFCs are already banned. The purpose is to diminish the placing of these refrigerants on the market by 79% from the volume of 2015 level by 2030, as shown in Figure 1. Please note that HFCs are very common refrigerants by now, of which R410a, R407c, and R134a are some examples. That being said, this regulation affects strongly and widely the refrigeration industry, and so the research and development process.

The refrigerant regulation is the reason for a tremendous shift of popular HFCs to more environmental-friendly refrigerants in the coming years. At the moment, HFOs seem to be suitable replacements because they have zero ODP and very low GWP. Also, they have compatible working pressures to common HFCs and good thermodynamic properties. The fact that they have just recently been studied for the last ten years, and there are only several studies on ERS using these refrigerants, which used the drop-in replacement method, as mentioned before, having a reasonable approach to the topic "HFOs for ERS" can assure a favorable performance and stability system.

The objective of the present work was the assessment of two hydrofluoroolefins for the ejector cooling technology using a mathematical method. Aiming to an ejector air-conditioning system that works with a low-grade heat source, the working temperature ranges were chosen as follows: the generator temperature was from 65 to 95 °C, the condenser temperature was fixed at 30 °C. The temperature at the condenser was 10 °C since this ERS was used for air conditioning purposes. Working fluids' assessment was based on several important parameters, such as the COP as a function of T_g and T_c, the required working pressure in a range of the generator temperature.

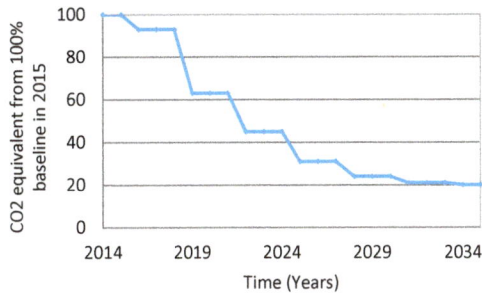

Figure 1. Expected phasing down process according to EU regulation 517/2014.

2. Mathematical Model of ECT

Referring to the basic ejector refrigeration cycle in Figure 2, the system consists of a vapor generator (boiler), refrigerant pump, ejector, condenser, evaporator, and expansion valve. There are three auxiliary cycles for maintaining a generator, condenser, and evaporator at desired temperatures. The ejector's working principle was described in works [13,14]. Please note that the letters in Figure 2 denote for the thermodynamic state of the refrigerant at the outlet of each component, (G) for generator, (C) for condenser, (D) for diffuser, (E) for evaporator, (V) for expansion valve, (P) for refrigerant pump.

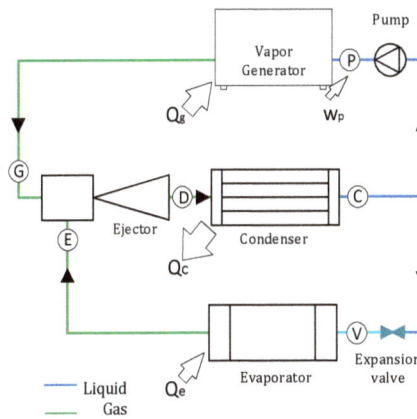

Figure 2. The basic working cycle of an ejector refrigeration system.

The coefficient of performance, COP, is the system performance. It is the ratio of evaporator heat energy Q_e to the total energy used to drive the cycle, the generator heat energy Q_g and the work of the pump, W_p

$$COP = \frac{Q_e}{Q_g + W_p} \tag{1}$$

Usually, W_p is negligible when compared to Q_e and Q_g. Hence, the above equation can be rewritten in the form of rates of heat flow:

$$COP \cong \frac{\dot{Q}_e}{\dot{Q}_g} = \frac{\dot{m}_e(h_e - h_v)}{\dot{m}_g(h_g - h_p)} \tag{2}$$

The cooling capacity, CC, is then defined as

$$\dot{Q}_e = \dot{m}_e(h_e - h_v) \tag{3}$$

The ejector distributes the key performance factor, the entrainment ratio (ER), to the whole ejector cooling system. Therefore, the mathematical model of ERS was focused on the ejector. The constant-pressure mixing ejector generally has better performance than constant-area mixing ejector, as confirmed in [11,15]. Constant-pressure mixing ejector is an ejector that has the nozzle exit position (NXP) within the suction area (convergent part), in front of the constant area, as shown in Figure 3.

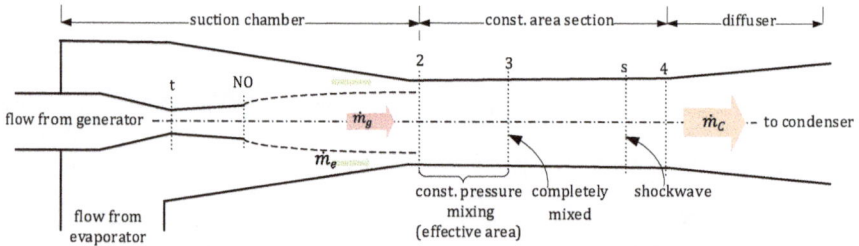

Figure 3. Schematic view of constant-pressure mixing ejector.

The following assumptions were made to simplify the model: (i) Thermodynamic processes are adiabatic, except for heat exchangers, i.e., generator, condenser, evaporator. (ii) Flow inside the system is isentropic, with accounting coefficients regarding irreversible processes, frictions; these coefficients were obtained from literature and experimental work. (iii) There is a section within the mixing chamber, where the secondary flow reaches a velocity of sound (when the ejector, or generally ERS is at the critical mode); this section is called hypothetical throat, as proposed by Munday and Bagster [13]. (iv) Primary and secondary flow are mixed at the hypothetical throat (cross-section 2); before section 2, they are separated flows, and after, they are completely mixed. (v) Velocities at two inlets and outlets of the ejector are neglected.

As seen in Figure 3, the ejector is divided into several sections. For each section, conservation of mass, conservation of momentum, and conservation of energy are applied to define the state of that section. Thermodynamic properties of working fluids are called from the real gas database of the program Engineering Equation Solver (EES). Following text presents relations of one block to other section by section:

The thermodynamic model of the ejector was governed based on the work of Huang et al. [16] and several assumptions from the work of Varga et al. [14] and Munday and Bagster [13].

The primary flow from the generator was assumed to have some degrees of superheat; the reason is superheated working fluid at the ejector will be in the form of gas (one phase ejector).

The working fluid inside the ejector is an ideal gas.

$$\dot{m}_g = A_t \frac{p_g}{\sqrt{T_g + T_{sup}}} \sqrt{\eta_t \frac{\kappa}{R} \left(\frac{2}{\kappa+1}\right)^{\frac{\kappa+1}{\kappa-1}}}, \tag{4}$$

where p_g is the pressure of the primary flow. T_g is the saturated temperature at pressure p_g. T_{sup} is the degree of superheating of primary flow. η_t is the isentropic coefficient of the primary nozzle. κ is the ratio of specific heats of given fluid, and R is the universal gas constant.

The relation between the nozzle outlet area A_{NO} and throat area A_t is defined from the gas dynamic relation of Mach number at the nozzle outlet Ma_{NO}

$$A_{NO} = \left(\frac{A_t}{Ma_{NO}}\right)\left[\frac{2}{\kappa+1}\left(1+\frac{\kappa-1}{2}Ma_{NO}^2\right)\right]^{\frac{(\kappa+1)}{2}} \tag{5}$$

The pressure at the nozzle outlet is defined by the total pressure of the primary flow and Mach number at the nozzle outlet by the following equation

$$p_{NO} = p_g\left(1+\frac{\kappa-1}{2}Ma_{NO}^2\right)^{\frac{-\kappa}{\kappa-1}}. \tag{6}$$

When the system works at the critical point (see Figure 4), the static pressure somewhere right behind the primary nozzle outlet p_{NO} is approximately equal to the pressure at the secondary inlet p_e. For simplification, Varga et al. [14] assumed that:

$$p_{NO} \cong p_e. \tag{7}$$

For a given working fluid, the pressure at the secondary inlet to ejector is a function of evaporator temperature, x is vapor fraction, "x = 1" indicates that the working fluid is at saturated gas state:

$$p_e = pressure(T = T_e, x = 1). \tag{8}$$

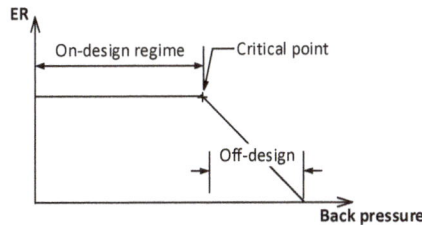

Figure 4. Illustration of various working regimes of an ejector.

According to Munday and Bagster [13], the hypothetical throat is the cross-section where the secondary flow chokes (reaches sonic condition) at the mixing chamber when the system works at critical mode. Huang et al. [16] later assumed that the choke takes place at cross-section "2", within the constant area section, as shown in Figure 2. Thus, Mach number of secondary flow at the hypothetical throat (Ma_{e2}) is assumed to be 1,

$$Ma_{e2} = 1. \tag{9}$$

The pressure of the secondary flow at the hypothetical throat p_{e2} is related with pressure at the secondary inlet p_e by following dynamic function for isentropic flow

$$p_{e2} = p_e\left(1+\frac{\kappa-1}{2}Ma_{e2}^2\right)^{\frac{-\kappa}{\kappa-1}}. \tag{10}$$

At the hypothetical throat, the pressure of primary flow p_{g2} is equal to the one of secondary flow p_{e2}, according to [13]

$$p_{g2} = p_{e2}. \tag{11}$$

For a complete set of equations of the ejector, the reader is referred to the work of Huang et al. [16].

Assuming that the pressure drop between diffuser and condenser is negligible, the rate of heat flow that is needed to transfer from the ejector cycle is,

$$\dot{Q}_c = \left(\dot{m}_g + \dot{m}_e\right)(h_c - h_d),\tag{12}$$

where h_d and h_c are the enthalpy at the outlet of ejector and condenser, respectively.

The heat loss during the heat transfer between the ejector cycle and the auxiliary cycle at the condenser is counted by the coefficient η_c.

η_c varies by the ambient temperature, insulation, etc. Thus, it will be collated with results from experimental work.

The process in the expansion valve (V) was assumed as an adiabatic process. Thus, $h_c = h_v$. Also, it was assumed that heat loss at the evaporator is negligible because the evaporator is well insulated and the heat flow is continuous and steady. The cooling capacity is related to the secondary flow rate as the following equation

$$\dot{Q}_e = \dot{m}_e(h_e - h_v)\tag{13}$$

The pump power used to lift the pressure from the condenser pressure level to generator pressure (assuming that pressure from the outlet of the pump is equal to the pressure at inlet generator) is,

$$\dot{W}_p = \frac{\dot{m}_g\left(p_g - p_c\right)}{\rho\cdot\eta_p},\tag{14}$$

where the density ρ is a function of condenser temperature and pressure because it is the inlet valve of the pump. The coefficient at the pump η_p is used to count on friction loss.

The required rate of heat flow at the generator is then calculated by

$$\dot{Q}_g = \frac{\dot{m}_g\left(h_g - h_p\right)}{\eta_g},\tag{15}$$

where the generator coefficient η_g is used to compute heat loss during the heat transfer process from the solar collector cycle to the ejector cycle. The enthalpy at the pump outlet h_p is the sum of the enthalpy at the condenser outlet and the enthalpy generated by the pump

$$h_p = h_c + \frac{\dot{W}_p}{\dot{m}_g}.\tag{16}$$

Thermodynamic properties at each stage of the whole cycle are called from the real gas database.

3. Refrigerant Selection

Working fluids should satisfy the criteria for system performance, environmental safety, economics. Thermodynamic properties of refrigerants are the most cared factors by researchers when choosing working fluid for the ERS. They must have low environmental impacts, meaning low GWP and ODP. They should be non-toxic, non-corrosive, chemically stable, and non-explosive.

In this study, six refrigerants were selected for the assessment. Some common HFCs were selected as references for comparison—R32 and R152a. They have an acceptable GWP index, inexpensive, non-toxic. Therefore, it is recommended as a substitution for refrigeration systems in the refrigerant regulation 517/2014. R600a and R290 are natural refrigerants; they have favorable thermodynamic properties and low environmental impact. Nevertheless, they are flammable. Two HFOs, R1234ze(e) and R1234yf, are the fourth-generation refrigerants of halocarbons. Their most valuable properties are very low GWP and non-flammability.

Table 1 presents the important properties of the selected refrigerants. These properties directly influence the performance of the refrigerants and the use of them in regard to the refrigerant regulation. Although the ozone depletion potential (ODP) is one of the most important parameters for refrigerant qualification, it is not included in the table because all these working fluids have zero ODP. The other factors are shortly explained in the following text:

- Molecular weight M_{mol} is an important factor that influences the system size at a given cooling capacity (CC) and working conditions. For a set of working conditions, the CC depends on $\dot{m}_e$ and Δh (see Equation (13)). Assume that the difference of Δh between working fluid is negligible, then the CC is a function of $\dot{m}_e$ only. Consider the equation of state of ideal gas below:

$$pV = \frac{m}{M}RT, \tag{17}$$

where V is the volume of working fluid in the gas state, M is molecular weight. Parameters p, R, and T are constants as the assumptions. Accordingly, the higher the M is, the lower V is required. In other words, the system is more compact if the molecular weight of refrigerant is large. As it is shown in Table 1, using HFOs as working fluid has a number of valuable aspects, such as:

- GWP of carbon dioxide is equal to 1 as it is chosen as the reference gas for defining GWP. GWP100 of gas tells us how much heat the gas traps in the atmosphere in 100 years compared to carbon dioxide.
- Safety class: According to the ASHRAE standard, hazardous is presented by a capital letter and a number. Toxicity is indicated by a capital letter: letter A—low toxicity, and B, high toxicity. Flammability is classified into four groups: Group 1—non-flammable, group 2—mildly-flammable, group 2L—lower-flammable, and group 3—high-flammable. R1234ze(e) and R1234yf are classified as lower-flammable; which is only considerable drawback of these HFOs.
- SSL is the Slope of the Saturated-vapor Line in the temperature-entropy (T-s) diagram. Isentropic expansion of a working fluid that has a negative slope of the saturated-vapor line may lead to a condensation [16], which will affect the performance of the ejector. To avoid this issue, some degrees of superheat should be applied. The need for superheating depends on how negative the slope is. With a positive-slope refrigerant, the superheating is necessary as well, but with less intensity. For example, superheating was applied in an ERS with R600a, which has a positive-slope of saturated line, as a working fluid in the works of Varga et al. [17,18].
- L_{ratio} of a working fluid is the latent heat ratio of that working fluid. It is defined by the following equation:

$$L_{ratio} = \frac{h_{vap,e} - h_{liq,e}}{h_{vap,g} - h_{liq,g}}, \tag{18}$$

where $h_{vap,e}$ and $h_{liq,e}$ are enthalpy of saturated vapor and saturated liquid at the evaporator temperature. Similarly, $h_{vap,g}$ and $h_{liq,g}$ are enthalpy of saturated vapor and saturated liquid at the generator vapor, respectively.

- The critical pressure (p_{crit}) of the critical point influences the required robustness of the system. The ejector cooling system obtains a preferable performance if the temperature of the driving heat source (the primary flow) close to the critical temperature (T_{crit}). Which implies that the critical pressure is mostly equivalent to the working pressure of the primary flow. A low critical-pressure refrigerant requires a less system robustness and a lower pressure-head pump; and thus, it is desirable.

Table 1. Key properties of selected refrigerants.

	M_{mol} (g/mol)	T_{cri} (°C)	P_{cri} (bar)	GWP100	Safety Class	SSL	L_{ratio}
R32	52	78.4	53.8	675	A2L	Negative	2.34
R152a	66.1	113	45.2	124	A2	Negative	1.56
R600a	58.1	135	36.3	3	A3	Positive	1.48
R290	44.1	96.7	42.5	20	A3	Negative	1.9
R1234yf	114	94.7	33.8	4	A2L	Positive	2
R1234ze(e)	114	109	36.4	7	A2L	Positive	1.9

4. Validation of the Mathematical Model

The mathematical model was validated with experimental results. The experimental works were implemented in a solar ejector refrigeration system with R600a as the working fluid. System calibrations and uncertainty analyses were carried out to ensure the reliability of the experimental results. For more details, the reader is referred to the work of Varga et al. [17,19].

As seen in Figure 5, the predicted values from the mathematical model well agreed with the experimental results, values of both entrainment ratio (ER) and the coefficient of performance (COP) were within ±15%. Apparently, the superheating of primary and secondary inlet flows significantly influenced the system performance, as well as the system accuracy. From the experimental data, both flows usually were in the state of a high degree of superheats, up to 15 K. However, in the mathematical model of this study, only the superheating of the primary inlet (stream from the generator) was considered.

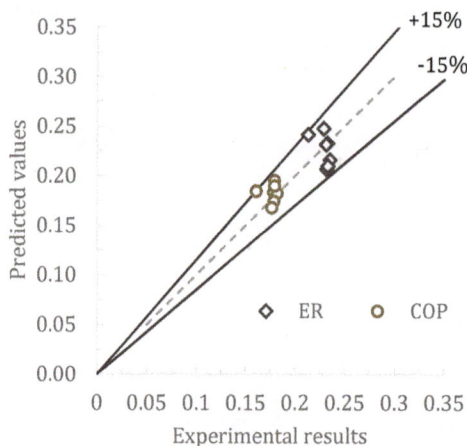

Figure 5. Validation of mathematical model by experimental results.

5. Results and Discussion

5.1. The Entrainment Ratio and the Performance of the System

The entrainment ratio (ER) as a function of generator temperature is presented in Figure 6. At the constant temperature of the condenser (30 °C) and evaporator (10 °C), the ERs increased with the generator temperature. At the T_g of 65 °C, entrainment ratios of all selected refrigerants reached around 25%, except R32. R32 was recommended as an alternative by regulation 517/2014 due to its availability and reasonable low GWP index. However, it did not have a desirable performance in ECT. ERs of R600a and R1234ze(e) were quite comparable with each other, and they were the highest among all others. ER of R1234yf was significantly higher than of R290, but their profiles with the change of generator temperature were very similar. Increasing further the T_g to over 95 °C apparently did not improve the ERs of these two refrigerants at all.

The performance of the system at various generator temperatures is shown in Figure 7. The system COP is a product of ER and the ratio of enthalpy differences of the phase-change processes (via Equation (2)). As seen, the performance of R152a was the highest among all, while its ER was just an average value compared to others. A similar observation was found for the performance of R32, which was comparable to the performance of R600a. The reason was the latent heat that used to evaporate the liquid to the gas state at the evaporator (low temperature) was higher than the latent heat that used to change the refrigerant from liquid state to gas state at the generator (high temperature). When the T_g approached greater values (closer to critical temperatures), the required latent heat at the generator became even smaller. Thus, the denominator of Equation (2) became smaller, resulting in a higher COP. This observation gave us a thought that having a generator temperature close to the refrigerant's critical temperature might yield better performance.

The performance of R1234ze(e) in the whole range of T_g was favorable, it reached 45% at the T_g of 95 °C. On the other hand, the COP of R600a was noticeably lower than the COP of R1234ze(e) due to the difference of latent heats of the evaporating process. Even though R1234ze(e) and R1234yf are isomers, the performance of R1234yf was clearly poorer than the other.

As a consequence, the required rate of heat flow at the generator $\dot{Q}_g$ for a fix cooling capacity of the ERS was a function of COP, as presented in Equation (2).

When required cooling capacity $\dot{Q}_e$ is constant, the change of COP is inversely proportional to $\dot{Q}_g$. For each refrigerant, with a constant T_c and T_e (also constant P_c and P_e), the COP generally is relative to T_g and P_g. Therefore, low T_g means low COP, and low COP requires higher $\dot{Q}_g$. In other words, in order to conduct a required entrained flow $\dot{m}_e$, higher primary flow $\dot{m}_g$ is required when the primary temperature is low. It means that a larger ERS is required if we want to work with a "low-grade driven heat source", thus higher investment is required. For example, the required $\dot{Q}_g$ of the 4 kW cooling capacity ERS with R600a as working fluid at T_g of 60 °C (with 5K of superheat) was 33.5 kW; the corresponding value when $T_g = 80$ °C (with 5K of superheat) was 12.1 kW, almost 3 times lower $\dot{Q}_g$ was needed with 20 K difference of primary working temperature.

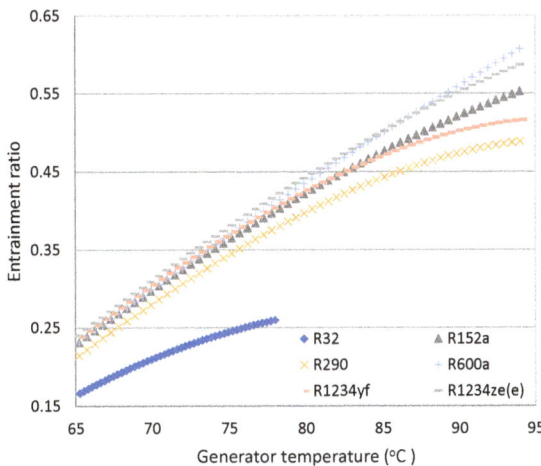

Figure 6. Entrainment ratio as a function of generator temperature ($T_e = 10$ °C and $T_c = 30$ °C).

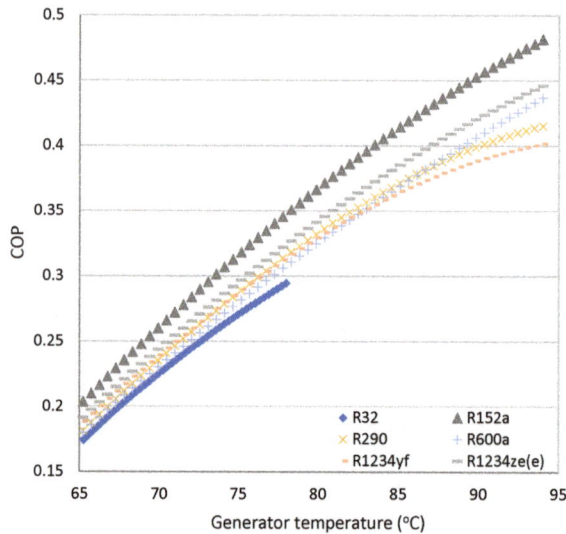

Figure 7. System performance as a function of generator temperature ($T_e = 10\,°\text{C}$ and $T_c = 30\,°\text{C}$).

5.2. Working Pressure

The working pressure is the most concerning factor in designing a refrigeration system generally. A system with high working pressure requires robust construction and durable equipment. Particularly, this statement is compelling for ERSs. The refrigerant pump is an example; typically, a refrigerant pump in an ERS is required to generate high working pressures at low flow rates. Such a pump is not easy to find in the market at the moment, especially the one that can generate a pressure of 20 bar and higher. In the scope of this paper, the pressure level at the primary inlet flow would be discussed as it is the highest, and thus, is the most significant to the system in the viewpoint of robustness. The primary pressures were defined from the required primary inlet temperatures, which are working conditions, inputs of the mathematical model.

As expected, the required primary pressures of the working fluids varied widely and increased with the generator temperature (via Figure 8). The R600a showed its advantage in this aspect; the primary working pressure varied only from 9 bar to 18 bar as the T_g reached 95 °C. This was one of the reasons for many researchers to choose isobutane in their experimental systems, as discussed previously. The p_g of R1234ze(e) was reasonably good, in the range from 13 to 27 bar, about 33% higher than R600a. As mentioned earlier, a high-pressure range raises the demand for the system's robustness and durability, such as a generator, condenser. These heat exchangers in an ERS were usually plate and frame compact heat exchangers, which have limits in handling high-pressure fluids.

R152a worked at a pressure range from 15 to 31 bar at the saturated temperature from 65 to 95 °C, roughly 68% higher than the R600a. R1234yf required a higher-pressure range than R152a by about 1.5 bar. The ERS system using R152a and R1234yf can be realized at the high initial cost due to the pressure limit of the commercial plate-and-frame heat exchanger.

R290 and R32, on the other hand, required very high working pressures, which most likely exceed the pressure range of a typical plate-and-frame compact heat exchanger. Therefore, such ERS is technically and economically very challenging, especially for the low-performance refrigerant like R32.

5.3. Area Ratio

The area ratio (AR) is the ratio of the mixing area A_2 over the throat area of the primary nozzle A_t

$$AR = \frac{A_2}{A_t}. \tag{20}$$

As mentioned, ejector geometry was defined based on the assumption that the secondary flow at the mixing chamber (sections 2–3 in Figure 3) reaches sonic speed, and the primary flow at the outlet of the primary nozzle is neither over expansion nor under expansion flow but column-shaped. This means the nozzle works at the optimal state for given conditions, producing maximum efficiency.

The variation of AR presents the sensitivity of the optimal geometry at various working conditions. The higher the differences of AR, the worst the system performs when the actual operating conditions are different than the designed conditions. As shown in Figure 9, the AR of R32 varied from 1.9 to 2.5 when the T_g went from 65 to 77.7 °C. We could say that the performance of R32 did not vary strongly when the generator temperature, or working conditions in general, varied from the designed (optimal) dimensions. On the contrary, the AR of R600a and R1234ze(e) varied from 3 to 7 when the generator temperature changed from 65 to 95 °C. It implies that the system would be sensitive to the actual operating conditions, no matter what the optimally designed geometry might be. Therefore, the system could be considered as less stable.

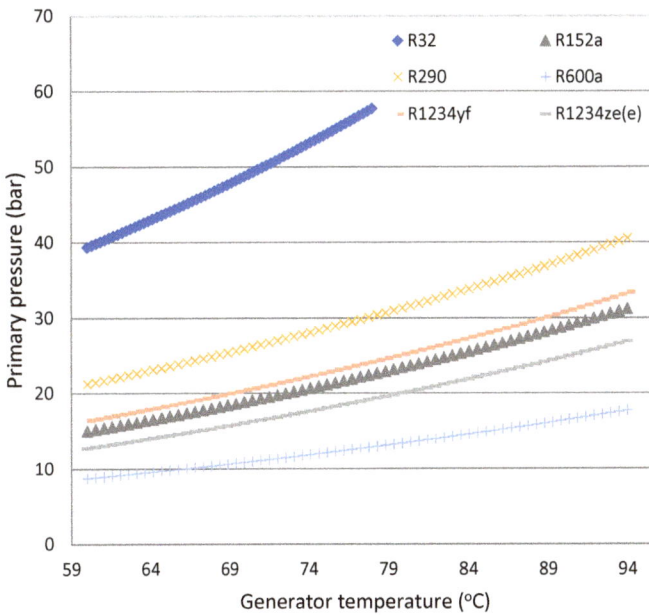

Figure 8. Working pressure of refrigerants at various generator temperatures.

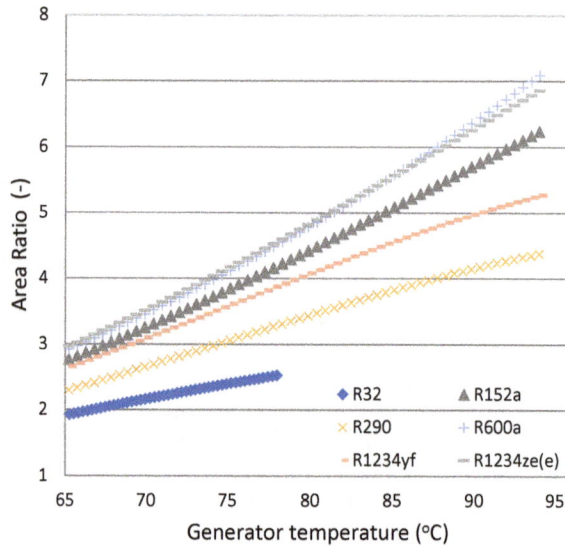

Figure 9. The area ratio as the function of generator temperature.

6. Conclusions

In this paper, the six selected working fluids were assessed based on three essential criteria: the coefficient of performance, the working pressure, and the sensitivity of optimal ejector geometry with working conditions.

The used mathematical model employed several coefficients to compensate for the isentropic and friction losses in the ejector cycle. The results indicated reasonable agreement when compared with a numerical model. More detail on model validation will be presented in our next study.

Overall, the HFC R152a and the HFO R1234ze(e) were the best performance refrigerants. The hydrocarbon R600a also offered promising performance. However, the flammability and the high-GWP-value of the hydrocarbon and HFC made HFO R1234ze(e) the only suitable choice.

Due to the diversity of working pressures, the required robustness of ERS significantly varies by different working fluids. The selections of the working pump and heat exchangers must be based on selected working fluids in this sense. The assessment on working pressure also showed that the three working fluids that have just been mentioned were best ones, from which R600a was the most favorable. R1234ze(e) also proved that it is a great candidate for ERS as well as for the refrigeration system in general. However, neither R152a nor R1234ze(e) showed the desired character in terms of system stability with a fixed ejector geometry. The solution for the issue is using a variable geometry, which is adjustable so that it can reach (or close to) optimal geometry for actual operating conditions.

Author Contributions: Conceptualization, V.V.N. and S.V.; methodology, S.V. and V.V.N.; software, S.V. and V.V.N.; validation, V.V.N., S.V. and V.D.; formal analysis, V.V.N. and S.V.; investigation, V.V.N. and S.V.; resources, S.V.; data curation, V.V.N. and S.V.; writing—original draft preparation, V.V.N.; writing—review and editing, S.V. and V.D.; visualization, V.V.N. and S.V.; supervision, S.V. and V.D.; project administration, S.V. and V.D.; funding acquisition, V.D. and V.V.N.

Funding: This publication was written at the Technical University of Liberec as part of the project 21226/2220 and the project "Experimental, theoretical and numerical research in fluid mechanics and thermomechanics", no. 21291 with the support of the Specific University Research Grant, as provided by the Ministry of Education, Youth and Sports of the Czech Republic in the year 2019.

Conflicts of Interest: The authors declare no conflict of interest. The funders had no role in the design of the study; in the collection, analyses, or interpretation of data; in the writing of the manuscript, or in the decision to publish the results.

Nomenclature

$\dot{Q}$	heat rate (kW)
$\dot{V}$	volumetric flow rate (ms^{-1})
$\dot{W}$	work (kW)
$\dot{m}$	mass flow rate (gs^{-1})
A	area (m^2)
AR	area ratio (-)
CC	cooling capacity (kW)
COP	coefficient of performance (-)
ER	entrainment ratio (-)
h	specific enthalpy (kJkg^{-1})
Ma	Mach number (-)
p	absolute pressure (bara)
P	relative pressure (bar)
R	universal gas constant J/(mol·K)
T	temperature (°C)
η	coefficient (-)
κ	specific heat ratio, (-)
ρ	density (kg m^{-3})

Subscripts

C	condenser
crit	critical
D	diffusor
E	evaporator
G	generator
in	inlet
n	nominal
NO	Nozzle outlet
out	outlet
P	pump
sat	saturated
sup	superheating
t	throat
th	thermal
V	valve

References

1. Guo, J.; Shen, H. Modeling solar-driven ejector refrigeration system offering air conditioning for office buildings. *Energy Build.* **2009**, *41*, 175–181. [CrossRef]
2. Huang, B.-J.; Chang, J.; Petrenko, V.; Zhuk, K. A solar ejector cooling system using refrigerant R141b. *Sol. Energy* **1998**, *64*, 223–226. [CrossRef]
3. Khalil, A.; Fatouh, M.; Elgendy, E. Ejector design and theoretical study of R134a ejector refrigeration cycle. *Int. J. Refrig.* **2011**, *34*, 1684–1698. [CrossRef]
4. Yu, J.; Li, Y. A theoretical study of a novel regenerative ejector refrigeration cycle. *Int. J. Refrig.* **2007**, *30*, 464–470. [CrossRef]
5. The European Parliament. *Regulation (EU) No 517/2014 of the European Parliament and of the Council*; The European Parliament: Brussels, Belgium, 2014.
6. Fang, Y.; Croquer, S.; Poncet, S.; Aidoun, Z.; Bartosiewicz, Y. Drop-in replacement in a R134 ejector refrigeration cycle by HFO refrigerants. *Int. J. Refrig.* **2017**, *77*, 87–98. [CrossRef]
7. Śmierciew, K.; Gagan, J.; Butrymowicz, D.; Łukaszuk, M.; Kubiczek, H. Experimental investigation of the first prototype ejector refrigeration system with HFO-1234ze(E). *Appl. Therm. Eng.* **2017**, *110*, 115–125. [CrossRef]
8. Śmierciew, K.; Gagan, J.; Butrymowicz, D.; Karwacki, J. Experimental investigations of solar driven ejector air-conditioning system. *Energy Build.* **2014**, *80*, 260–267. [CrossRef]

9. Roman, R.; Hernandez, J.I. Performance of ejector cooling systems using low ecological impact refrigerants. *Int. J. Refrig.* **2011**, *34*, 1707–1716. [CrossRef]
10. Sankarlal, T.; Mani, A. Experimental investigations on ejector refrigeration system with ammonia. *Renew. Energy* **2007**, *32*, 1403–1413. [CrossRef]
11. Besagni, G.; Mereu, R.; Inzoli, F. Ejector refrigeration: A comprehensive review. *Renew. Sustain. Energy Rev.* **2016**, *53*, 373–407. [CrossRef]
12. Yu, J.; Du, Z. Theoretical study of a transcritical ejector refrigeration cycle with refrigerant R143a. *Renew. Energy* **2010**, *35*, 2034–2039. [CrossRef]
13. Munday, J.T.; Bagster, D.F. A New Ejector Theory Applied to Steam Jet Refrigeration. *Ind. Eng. Chem. Process. Des. Dev.* **1977**, *16*, 442–449. [CrossRef]
14. Varga, S.; Lebre, P.S.; Oliveira, A.C. Readdressing working fluid selection with a view to designing a variable geometry ejector. *Int. J. Low-Carbon Technol.* **2013**, *10*, 205–215. [CrossRef]
15. Tashtoush, B.; Alshare, A.; Al-Rifai, S. Performance study of ejector cooling cycle at critical mode under superheated primary flow. *Energy Convers. Manag.* **2015**, *94*, 300–331. [CrossRef]
16. Huang, B.-J.; Chang, J.; Wang, C.; Petrenko, V. A 1-D analysis of ejector performance. *Int. J. Refrig.* **1999**, *22*, 354–364. [CrossRef]
17. Varga, S.; Oliveira, A.C.; Palmero-Marrero, A.; Vrba, J. Preliminary experimental results with a solar driven ejector air conditioner in Portugal. *Renew. Energy* **2017**, *109*, 83–92. [CrossRef]
18. Pereira, P.R.; Varga, S.; Oliveira, A.C.; Soares, J. Development and Performance of an Advanced Ejector Cooling System for a Sustainable Built Environment. *Front. Mech. Eng.* **2015**, *1*, 1–12. [CrossRef]
19. Pereira, P.R.; Varga, S.; Soares, J.; Oliveira, A.C.; Lopes, A.M.; De Almeida, F.G.; Carneiro, J.F. Experimental results with a variable geometry ejector using R600a as working fluid. *Int. J. Refrig.* **2014**, *46*, 77–85. [CrossRef]

energies

Article

Looking for Energy Losses of a Rotary Permanent Magnet Magnetic Refrigerator to Optimize Its Performances

Angelo Maiorino [1,*], Antongiulio Mauro [1], Manuel Gesù Del Duca [1], Adrián Mota-Babiloni [2] and Ciro Aprea [1]

[1] Department of Industrial Engineering, Università di Salerno, Via Giovanni Paolo II, 132, Fisciano, 84084 Salerno, Italy; amauro@unisa.it (A.M.); mdelduca@unisa.it (M.G.D.D.); aprea@unisa.it (C.A.)
[2] ISTENER Research Group, Department of Mechanical Engineering and Construction, Campus de Riu Sec s/n, Universitat Jaume I, E-12071 Castelló de la Plana, Spain; mota@uji.es
* Correspondence: amaiorino@unisa.it; Tel.: +39-(0)-89-964105

Received: 27 September 2019; Accepted: 14 November 2019; Published: 19 November 2019

Abstract: In this paper, an extensive study on the energy losses of a magnetic refrigerator prototype developed at University of Salerno, named '8MAG', is carried out with the aim to improve the performance of such a system. The design details of '8MAG' evidences both mechanical and thermal losses, which are mainly attributed to the eddy currents generation into the support of the regenerators (magnetocaloric wheel) and the parasitic heat load of the rotary valve. The latter component is fundamental since it imparts the direction of the heat transfer fluid distribution through the regenerators and it serves as a drive shaft for the magnetic assembly. The energy losses concerning eddy currents and parasitic heat load are evaluated by two uncoupled models, which are validated by experimental data obtained with different operating conditions. Then, the achievable coefficient of performance (COP) improvements of '8MAG' are estimated, showing that reducing eddy currents generation (by changing the material of the magnetocaloric wheel) and the parasitic heat load (enhancing the insulation of the rotary valve) can lead to increase the COP from 2.5 to 2.8 (+12.0%) and 3.0 (+20%), respectively, and to 3.3 (+32%), combining both improvements, with an hot source temperature of 22 °C and 2 K of temperature span.

Keywords: magnetic refrigeration; magneto-caloric effect; coefficient of performance; eddy currents; experimental; parasitic heat load; modelling

1. Introduction

Magnetic refrigeration is an emerging and environmental-friendly technology that uses a solid refrigerant exploiting the magneto-caloric effect (MCE), which is represented by a temperature change of the material when it is subjected to a change of an external magnetic field. Studies showed that this technology could lead to 20–30% energy savings compared to vapor compression refrigeration because of magnetization work recovery and lower entropy generation [1–4], as well as a reduction of the environmental impact of the refrigeration system [5,6]. Several prototypes have been designed and built so far, and their performances strictly depend both on the system design and operating conditions, as well as on the employed magnetocaloric material [7]. Although magnetic refrigeration is a promising technology to substitute vapor-compression systems, its performances are still lower than those provided by vapor-compression systems, regarding cooling power, temperature span, efficiency, and system design. Several apparatus, based on different constructive concepts, have been presented and widely characterized in the literature over the years [8–20]. A comprehensive overview of these experimental devices built so far can be found in different recent reviews [7,21].

Energy efficiency is one of the crucial characteristics to allow this technology becoming mature for markets, as well as to compare different system concepts. Several strategies to improve energy efficiency have been explored, whether focusing on theoretical aspects and thermodynamic cycles [22–24] or performing extensive studies on system energy performances [10,25–27], as well as coupling magnetic refrigerators with systems such as Stirling motors, geothermal probes, and ejectors [28–31]. Furthermore, different studies have been conducted on the optimal control of magnetic cooling devices to improve their performance, using both an experimental [32] and a modelling approach [33,34].

A rotary permanent magnet magnetic refrigerator, named '8MAG', was developed at University of Salerno [35] and first experimental data were carried out [15,16] concerning cooling power, temperature span and coefficient of performance (COP). The optimization of existing prototypes, in terms of design and performance, is currently a common subject in the magnetic refrigeration literature [36–41], showing COPs of the order of 5 with 5 K of temperature span and 0.5 with 25 K of temperature span. '8MAG' showed a maximum COP of 2.5 with 2 K of temperature span and an hot source temperature (T_H) equal to 22 °C [35]. Furthermore, '8MAG' showed a maximum second-law efficiency of 2.4% at T_H = 22 °C and a temperature span of 3.3 K. These results are comparable to the performance of the magnetic refrigerator prototype presented by Capovilla et al. [42].

Then, in the present work, several numerical analyses have been performed to estimate parasitic losses of '8MAG' and to identify a way for improving prototype performances. In detail, eddy currents generation and the parasitic thermal load were addressed. Identifying the possible energy losses can help to highlight the improvements which are needed to be made to a magnetic refrigerator in different operating conditions. Nevertheless, a detailed study is mandatory to characterize these energy losses and analyse their effect on energy performance. Some losses can be easily identified and quantified by a few experimental tests, such as friction losses, but others are more difficult to characterize, with several experiments required. Hence, to reduce the experimental efforts and generalize the results, the characterization of this kind of energy losses was performed by mathematical models, properly defined and validated.

2. The Prototype and the Experimental Measurement System

A detailed description of the experimental apparatus ('8MAG') was already provided in Aprea et al. [16]. However, it is needed to focus on some design details to identify the most relevant energy losses. '8MAG' is a rotative permanent magnet prototype equipped with two magnets disposed at 180 degree based on a double U configuration with about 1.2 T. Eight static regenerators, located in the air gap (43 mm) between poles of magnets and disposed in 45 degrees among them, are alternatively magnetized and demagnetized with the rotation of the magnets assembly. The regenerators are fixed on a frame, named magnetocaloric wheel (MCW), which is located inside the magnets gap and made of a diamagnetic aluminium alloy.

A rotary valve, coaxial with the magnetic assembly, imparts the direction of the heat transfer fluid (demineralized water) distribution through the regenerators. The rotary valve consists of two main parts: a stator (fixed to the MCW) and a rotor (connected to the drive shaft). The rotor acts as a rotary manifold and as a shaft for the magnetic system. The stator is divided into a hot sub-valve and a cold sub-valve and each sub-valve allows the connection between the regenerators and the rotary manifold. Each part of the rotary valve is made in stainless steel.

The prototype is shown in Figure 1 and it was widely described in Aprea et al. [16].

Several sensors were used to measure the most important variables to characterize the energy performance of the magnetic refrigerator prototype.

A torque meter has been used to measure mechanical torque, an encoder provides angular speed, calibrated resistance temperature detectors (RTD) four wires have been used for temperature measurements. Torque and temperature measurements have been carried out using a national instruments (NI) compactDAQ system and the NI LabVIEW software. Axial mechanical power has been measured using the torque meter and controlling the angular speed of a direct current (DC) brushless motor in closed loop through. The test apparatus is equipped with a 32-bit analog to digital (A/D) converter acquisition cards with sampling rate up to 10 kHz. In Table 1, a summary of the used instrumentation and relative accuracy are reported.

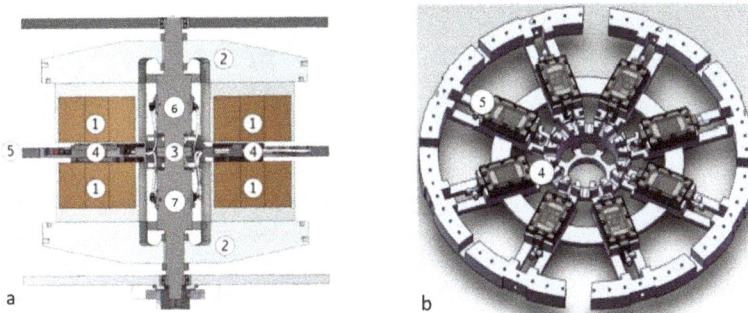

a

b

Figure 1. Prototype core details in cross-section (**a**) and 3-D view of MCW (**b**): (1) permanent magnet assembly; (2) magnets support; (3) shaft-rotary valve combination; (4) regenerators; (5) magneto caloric wheel (MCW); (6) cold sub-valve and (7) hot sub-valve.

Table 1. Measurement instruments used for the experimental tests.

Measurement	Instrument Type	Accuracy
Temperature	RTD 4 wires	0.1 K
Torque	Torque transducer	0.5%
Angular velocity	Optical encoder	$0.01° \, s^{-1}$
Magnetic field	Hall probe	0.4%
Water flow	Electromagnetic flowmeter	0.5%
Electrical power	Electromagnetic wattmeter	0.2%

3. Energy Losses Model

Considering the operating mode of '8MAG' and its design details, two different energy losses can be identified: mechanical losses and thermal losses. A core part of the entire assembly is the rotary valve, and therefore the overall study about energy losses can be performed focusing only on this component. Indeed, the rotary valve serves both as a drive shaft and as a thermal driver, distributing the cooling capacity and heat to be rejected. Hence, the model of energy losses, developed in COMSOL environment, is divided into two sub-models: the mechanical model and the thermal model. Both are graphically represented in Figure 2.

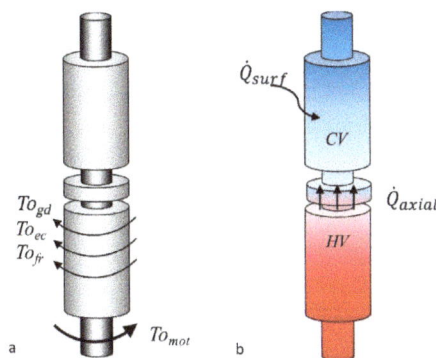

Figure 2. Representation of the mechanical model (**a**) and the thermal model (**b**).

3.1. Mechanical Model

The work needed to magnetize and demagnetize the regenerators depends on the rotation of the magnetic system by the rotary valve. Hence, a resistant torque (TO_{gd}) acts on this component, with a magnitude dependent on the working temperatures of the regenerators (hot and cold end). In detail, the magnitude of the resistant torque follows an oscillating trend, according to the alternative attraction and rejection of the magnets during the AMR cycle. This oscillation reduces with the increase of the rotational frequency due to the inertial phenomena related to the distribution of the rotating mass of the magnets. However, other two energy losses must be considered in '8MAG' related to the drive shaft, and therefore to the rotary valve.

The rotation of the magnetic system allows to magnetize and demagnetize the regenerators placed in the MCW. During the magnetization and demagnetization of the regenerators, also the MCW is subjected to the magnetic field, with the same intensity and frequency of the regenerators. This fact leads to generate eddy currents due to the electrical conductivity of aluminium. The occurrence of eddy currents causes a further resistant torque (TO_{ec}) that leads to increase the work needed to move the magnetic assembly. Moreover, the friction caused by the sliding of the bearings and seals of the rotary valve represents an additional resistant torque (TO_{fr}) on the drive shaft. Observing Figure 2, it is possible to write the following torque balance equation at steady-state conditions (neglecting the inertial term)

$$To_{tot}\left(\omega, T_{gd}, \beta\right) = To_{gd}\left(T_{gd}, \beta\right) + To_{ec}(\omega, \beta) + To_{fr}(\omega), \tag{1}$$

where β is the angular position of the magnets, T_{gd} is the working temperature of the regenerators, and ω is the rotational speed of the magnetic assembly, expressed in rotations per minute (rpm).

Then, the mechanical power balance, referred to a complete rotation, can be written as

$$\dot{W}_{tot}(\omega, T) = \dot{W}_{gd}(\omega, T) + \dot{W}_{ec}(\omega) + \dot{W}_{fr}(\omega), \tag{2}$$

where $\dot{W}_{ec}$ and $\dot{W}_{fr}$ represent the additional mechanical power required as a result of the mechanical losses ($\dot{W}_{loss}$). These latter two terms are considered in this work since they are source of losses that could be recovered. In Equation (2), the eddy currents generation in the magnetocaloric material and the air friction were neglected.

The evaluation of the effect of eddy currents (ECL—eddy currents losses), and then, the calculation of TO_{ec}, was performed by a mathematical model composed by three sub-model: the static magnetic field model (SMF), the stationary eddy currents power dissipation model (SECP), and the stationary thermal model (ST). The friction term ($\dot{W}_{fr}(\omega)$) was estimated by a semi-empirical approach, using technical data of the bearings and seals of the rotary valve and measuring the resistant torque ($TO_{fr}(\omega)$) to the rotary valve without the MCW and regenerators. Hence, it was possible to measure the resistant torque related to friction effects.

3.1.1. Static Magnetic Field Model

The evaluation of the magnetic field within the air gap, and therefore the intensity of the magnetic field to which the MCW is subjected, was carried out by a finite element method (FEM) analysis (see Figure 3). First, the domain under investigation was defined according to the real geometry of the magnetic assembly and the MCW. Then, a simplified geometry was designed to reduce computational time, neglecting holes, geometrical singularity, and low-relevant complex details. The entire geometry was included in a cylindrical volume with a diameter equal to 2.5 times of the diameter of the MCW and a height equal to 2.5 times of the height of the magnetic assembly. Three different meshes were tested: coarser (with 29,065 elements), normal (with 58,605 elements) and extremely fine (with 1,322,712 elements). A preliminary analysis of the standard deviation of the simulated magnetic flux density allowed to choose the best solution, which was the normal mesh with 58,605 elements.

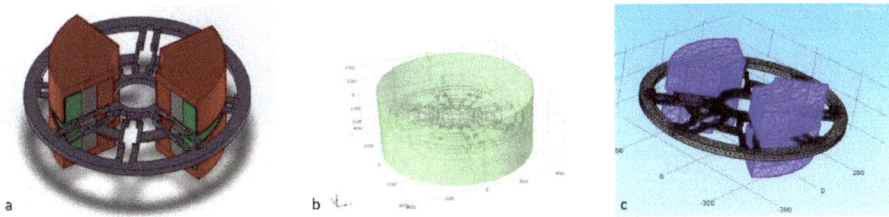

Figure 3. Modelled geometries (**a**), domain of the simulations (**b**), and meshes of magnets and the MCW (**c**).

The magnetic assembly was characterized considering the Halbach array configuration used in '8MAG' (see Figure 4), where each segment is made of sintered NdFeB with a magnetic remanence (B_r) of 1370 mT.

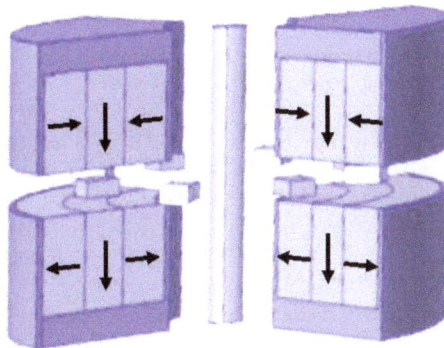

Figure 4. Configuration of the permanent magnets as Halbach array.

The model is defined by Maxwell's equation

$$\begin{aligned}
\nabla \times H &= J + \frac{\partial D_E}{\partial t}, \\
\nabla \times E &= -\frac{\partial B}{\partial t}, \\
\nabla \cdot D_E &= \rho_q, \\
\nabla \cdot B &= 0,
\end{aligned} \tag{3}$$

where H is the magnetic field intensity (in A m^{-1}), B is the magnetic flux density (T), E is the electric field intensity (V m^{-1}), D_E the electric flux density (C m^{-2}), J is the electric current density (A m^{-2}) and ρ_q is the volume charge density (C m^{-3}). The considered constitutive relations are shown in Equations (4)–(6)

$$D_E = \varepsilon_0 E + P, \tag{4}$$

$$B = \mu_0(H + M), \tag{5}$$

$$J = \sigma E, \tag{6}$$

where ε_0 is the vacuum permittivity (in F m^{-1}), P is the electric polarization (in C m^{-2}), μ_0 is the vacuum permeability (in H m^{-1}), M is the magnetization (in A m^{-1}) and σ is the electrical conductivity. The boundary conditions at the material interfaces and physical boundaries are represented by the equations

$$n_2 \times (E_1 - E_2) = 0, \tag{7}$$

$$n_2 \times (D_{E1} - D_{E2}) = \rho_s, \tag{8}$$

$$n_2 \times (H_1 - H_2) = J_s, \tag{9}$$

$$n_2 \times (B_1 - B_2) = 0. \tag{10}$$

At last, considering the current continuity Equation (Equation (11)), a further interface condition for the current density was introduced (Equation (12)).

$$\nabla \cdot J = -\frac{\partial \rho_q}{\partial t}, \tag{11}$$

$$n_2 \cdot (J_1 - J_2) = -\frac{\partial \rho_s}{\partial t}. \tag{12}$$

In Equations (8) and (12), ρ_s represents the surface charge density whereas, in Equation (9), J_s represents the surface current density. The SMF sub-model provides, as output, the magnetic field intensity and the distribution of the magnetic field within the investigated domain.

3.1.2. Stationary Eddy Currents Power Dissipation Model

The results of the SMF sub-model (intensity of the magnetic field and its distribution) represent the input of the second sub-model, that is the SECP model, which aims to evaluate eddy currents generation inside the MCW.

To achieve this target, a Lorentz type induced current density term is included in the previous equation set (Equations (11) and (12)). Magnetic field rotation is simulated supposing that the magnets are rotating at a constant rotational speed ω. Different steady-state simulations were performed at various relative magnets/wheel positions (β) simulating a complete rotation of the magnets. The effect of magnetic field variation in the MCW, due to the presence of regenerators, has been neglected, thus regenerators have not been included in the modelling.

The Lorentz term is related to the rotational speed of the magnets, as

$$v = \omega(-y, x, 0). \tag{13}$$

Then, the eddy currents dissipation, in terms of resistive losses, can be calculated as

$$q_{ec}(x, y, z) = \sigma J^2, \tag{14}$$

where σ is the electrical conductivity. The global power dissipated due to eddy currents is

$$Q_{ec} = \int \int \int_V \sigma J^2. \tag{15}$$

where V is the geometrical volume defined by the mesh. Results show both the typical vortices formation of induced currents in the metal because of the longitudinal magnetic field gradient. Furthermore, due to the lack of axial symmetry of the MCW geometry, eddy current dissipation results to be a function of the relative rotation β. Figure 5 shows the magnetic flux density in the MCW (Figure 5a–c), the induced current generated (red arrow in Figure 5b–d) and the specific power generated in the aluminium (Figure 5b–d) for two different angles β (0° and 25°).

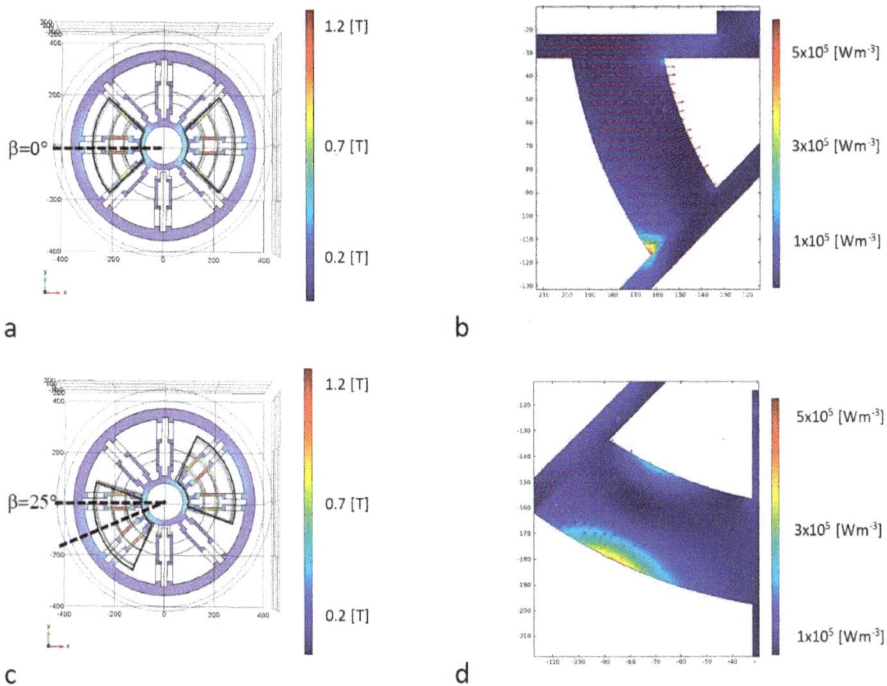

Figure 5. Magnetic flux distribution (**a**–**c**), induced current density (**b**–**d**), and resistive losses (**b**–**d**) for two different positions of the magnets.

3.1.3. Stationary Thermal Model

After evaluating the distribution of eddy currents on the MCW, a steady-state 3D thermal sub-model was implemented, starting from the resistive losses field as a heat generation term. It was assumed that convective and radiative heat exchange can be described by an equivalent constant convective coefficient term on the whole domain to simplify the computation. Indeed, the purpose of the simulation is to validate the order of magnitude of the resistive losses previously calculated. Even if the problem is not completely stationary due to the dynamic behaviour of the system, it was assumed that a global heat flux invests the MCW. This global heat flux is the spatial composition of different heat flux for various β values. Hence, different steady-state simulations were performed for different angular positions. A single position of the magnetic assembly and a single portion of MCW was considered to validate the model.

The heat transfer through the MCW is described by a steady-state energy balance equation, as

$$\nabla\cdot(-k\nabla T_{MCW}) = q_{ec}, \tag{16}$$

where the heat source (q_{ec}) is represented by the resistive losses due to eddy currents generation.

The heat dissipated on air-exposed surfaces is equal to

$$-n\cdot q = h(T_{air} - T_{MCW}), \tag{17}$$

where h is the equivalent heat transfer coefficient, supposed constant on the whole surface.

Since the magnets rotate, forced convection can be assumed depending on the magnets angular speed. In the analysed rotating frequency range, h resulted to be within 20 and 30 W m^{-2} K^{-1}, considering the following experimental correlation for the convective heat transfer coefficient (h_c) expressed in W m^{-2} K^{-1}, as a function of the rotating frequency f (in Hz)

$$h_c = 10.45 - f + 10f^{0.5}. \tag{18}$$

In Figure 6, the surface temperature of the MCW is shown for an ambient temperature of 20 °C and a rotating frequency equal to 0.72 Hz. The regenerator place mostly permeated by the magnetic field has been taken as reference for temperature increase validation.

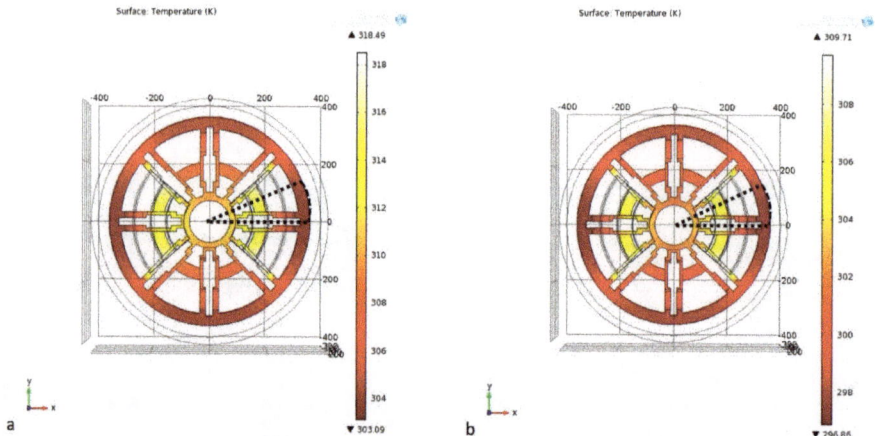

Figure 6. Temperature distribution on the MCW surface for a rotating frequency of 0.72 Hz, an ambient temperature of 20 °C and different heat transfer coefficients: 20 (**a**) and 30 (**b**) W m^{-2} K^{-1}. In the figures, the area of interest for the simulation is evidenced (black dotted lines).

The value of the power dissipated by generation of eddy currents can be assumed as

$$\dot{W}_{ec} = \dot{Q}_{ec}. \tag{19}$$

3.1.4. Semi-Empirical Evaluation of Friction Losses

The identification of the resistant torque concerning the friction effects into the rotary valve ($To_{fr}(\omega)$) was performed by a semi-empirical approach that allowed to identify the relation between the friction resistant losses and the operating frequency (f), expressed by Equation (20).

$$TO_{fr}(\omega) = 0.7586f + 4.7207 = 0.7586\frac{\omega}{60} + 4.7207. \tag{20}$$

In Figure 7, the results of the experimental measurements of the resistant torque without the MCW and the regenerators used to point out Equation (20) are shown.

The power dissipated due to friction effects was easily calculated as

$$\dot{W}_{fr}(\omega) = 2\pi f TO_{fr}(\omega). \tag{21}$$

However, the reduction of friction losses is not considered in this work as a possible improvement for '8MAG' since they are negligible compared to the energy losses concerning eddy currents generation.

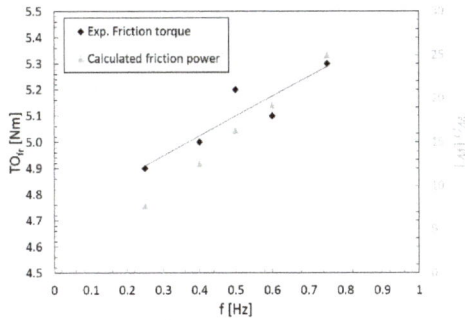

Figure 7. Experimental resistant torque by friction effects (left y-axis) and correlated power dissipated (right *y*-axis) for different operating frequency.

3.2. Thermal Model

The rotary valve is subjected to a heat exchange between its external surface and the surrounding air due to the temperature difference. Furthermore, an axial heat exchange occurs within the valve since the hot and cold sub-valve achieve different steady-state temperature levels during cyclic operations of '8MAG'. This temperature difference between the sub-valves is only related to the fluid flow through the internal ducts of the valve.

Considering the surface of the cold sub-valve, it is possible to identify an axial conductive thermal loss though the hot buffer of the valve ($\dot{Q}_{axial}$) and a convective radiative heat loss on the external surface ($\dot{Q}_{surf}$). Hence, the parasitic heat loss can be evaluated with the equation

$$\dot{Q}_{c,loss} = \dot{Q}_{surf} + \dot{Q}_{axial}. \tag{22}$$

A FEM analysis, based on a 3D steady-state thermal model, was performed to evaluate the intrinsic thermal load of the rotary valve. A 3D geometrical model of the rotary valve was developed whereas the internal water ducts were modelled as 1D channel thermally coupled to the geometry of the valve (see Figure 8).

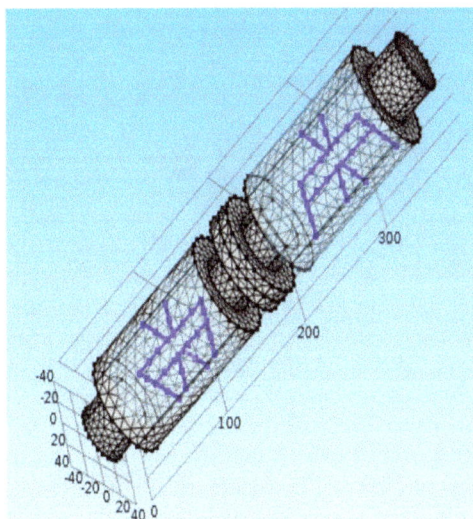

Figure 8. Representation of internal water ducts and the 3D geometry of the rotary valve.

As made for the mechanical model, the domain was first defined considering the real geometry of the rotary valve. Then, the geometry was simplified to reduce computational time, neglecting geometrical singularities and complex details of low relevance. The entire geometry was included in a cylindrical volume with a diameter equal to 2.5 times of the diameter of the rotary valve and a height equal to 2.5 times of the height of the entire valve. Three different mesh were testes: coarser (12,045 elements), normal (23,615 elements), and extremely fine (922,712 elements). Analysing the standard deviation of $\dot{Q}_{c,loss}$, the normal mesh was chosen as the best.

The following equations were solved for the 1D water flux

$$\nabla \cdot (A\rho u_w) = 0, \tag{23}$$

$$\nabla p + f_D \frac{\rho}{2D_h} u_w |u_w| = 0, \tag{24}$$

where u is the average surface velocity, ρ is the water density, p is the water pressure, D_h is the equivalent diameter, f_D is the Darcy friction coefficient, calculated with Churchill equation. In Equation (25), the energy balance equation is also shown as

$$\rho A c_p u_w \cdot \nabla T_w = \nabla \cdot A k \nabla T_w + f_D \frac{\rho A}{2D_h} |u_w| u_w{}^2 + \dot{Q}_{wall}, \tag{25}$$

where c_p is the specific heat at constant pressure, T_w is the water temperature and k is the thermal conductivity. $\dot{Q}_{wall}$ represents heat through pipe surface, which is equal to

$$\dot{Q}_{wall} = h_{int} Z \left(T_{ext} - T_w\right), \tag{26}$$

where Z is the perimeter of the duct, h_{int} is the internal heat exchange coefficient, T_{ext} is the external temperature of the pipe, which corresponds to the temperature of the solid (T_s).

The temperature field inside the valve is governed by the equation

$$\nabla \cdot (-k\nabla T_s) = 0, \tag{27}$$

where T_s is the solid temperature. The boundary conditions are represented by Equations (28) and (29)

$$-n \cdot (-k\nabla T_s) = h_{air}\left(T_{air} - T_{surf}\right), \tag{28}$$

$$-n \cdot (-k\nabla T_s) = 0, \tag{29}$$

where h_{air} is the convective heat transfer coefficient, T_{air} is the temperature of the air surrounding the rotary valve and T_{surf} is the valve surface temperature. End parts of the valve are assumed to be adiabatic. Solving the equations reported above, the thermal model allows to carry out the two components of $\dot{Q}_{c,loss}$, that are $\dot{Q}_{surf}$ and $\dot{Q}_{axial}$, and therefore to quantify the parasitic thermal losses in the rotary valve. The latter information is very useful since can help to understand which improvements can be made on the prototype to increase its overall performance.

4. Model Validation

The models introduced in Section 3 were validated by experimental tests for different working conditions, in terms of hot source temperature (T_H), operating frequency (f) and volumetric flow rate ($\dot{V}$). Only experimental tests at zero load were considered for model validation.

4.1. Mechanical Model Validation

To evaluate the performance of the mechanical model, it was required to validate only the sub-models used to calculate $\dot{W}_{ec}$ (SMF, SECP, and ST) since $\dot{W}_{fr}$ was estimated by a semi-empirical method (see Section 3.1.4).

The validation of the SMF sub-model was performed by carrying out some experimental measurements of the magnetic field by fixing a magnetometer on a radial axis and rotating step-by-step the magnets for different angular position β. The experimental results were then compared with the simulation in terms of the z component of the magnetic flux density B_z. A good agreement was found between experiments and simulation, highlighting that the model can reproduce magnetic field distribution, peaks, and valleys. Locally comparing measured and simulated data, a narrow absolute error emerged (-0.09 T), whereas the maximum relative error is of about 15%. The comparison between experiments and simulation of some representative points is shown in Table 2.

Table 2. Absolute and relative errors of the SMF model for some representative points.

X (cm)	y (cm)	B_{z_sim} (T)	B_{z_exp} (T)	Absolute Error (T)	Relative Error (%)
0	100	0.297	0.350	−0.053	−15.1
0	−180	−1.270	−1.180	−0.090	7.5
0	−200	−1.069	−1.085	0.020	−1.5
−100	−180	−0.827	−0.800	−0.030	3.4
−150	0	0.085	0.075	0.010	12.8

The calculated error values are acceptable considering the accuracy of the used instrumentation.

The SECP sub-model was validated measuring at different operating frequencies the resistant torque to the drive shaft with the MCW and removing all regenerators to exclude the term TO_{gd} from Equation (1). Hence, the measured torque is composed only by the friction term (TO_{fr}) and the eddy currents component (TO_{ec}). On the other hand, the related measured mechanical power represents the mechanical power losses $\dot{W}_{loss}$. The experimental data of resistant torque (indicated as TO_{loss}) and mechanical power ($\dot{W}_{loss}$) were used to make a comparison with the simulation data obtained by the SECP model. Since the SECP model does not consider friction effects, the simulation results, in terms of TO_{ec}, were increased with the friction losses contribution calculated with the semi-empirical model (see Equations (20) and (21)).

Hence, the comparison between the simulated and experimental resistant torque is shown in Figure 9, as well as the mechanical power loss. There is a good agreement between the measurements and the simulation results, with an average relative deviation of about 8.0% in the mechanical power (comparing the grey full line and grey symbols). This value is considered acceptable for the aim of this study. The greatest deviation in the mechanical power can be observed at $f = 0.75$ Hz. On the other hand, the simulated torque data (black full line) follow the trend of the experimental ones (black symbols) showing larger deviations for lower frequencies.

The ST model was validated comparing simulated data with experimental temperature measurements performed at different operating frequencies without regenerators and water flowing through the system. Six thermo-resistances, properly insulated, were placed on different representative positions on the metal surface of the MCW (see Figure 10).

Then, the surface temperature distribution of the MCW was obtained by the ST model and the simulated temperatures at the same position of Figure 10 were extracted. The comparison between the experimental and simulated temperature is shown in Figure 11.

The resulting value of the mean absolute percentage error (MAPE) is 4.0% with a maximum relative error of +7.8%, occurred for T3 (yellow line). These error values are considered acceptable for the aim of this work since are very close to the error range of the used instrumentation. The validation of the ST model demonstrates the capability of the mechanical model of providing a good estimation of the eddy current losses of '8MAG', which can be used to evaluate its achievable performance after their reduction.

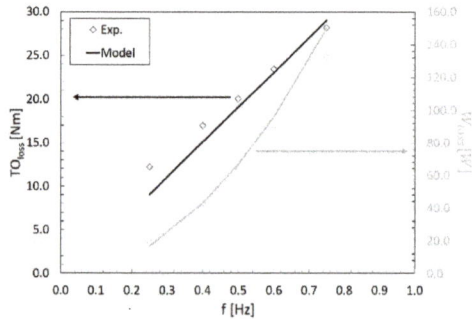

Figure 9. Comparison between experimental (symbols) and simulation (full line) results of resistant torque (black symbols and full line on the left *y*-axis, only losses contribution) and mechanical power losses (grey symbols and full line on the right *y*-axis).

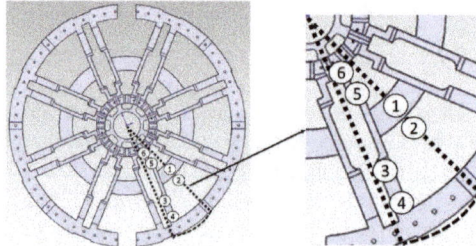

Figure 10. Temperature sensors positioning on the surface of the MCW. Sensors 1 and 2 were placed on the top and bottom surface of the MCW, respectively.

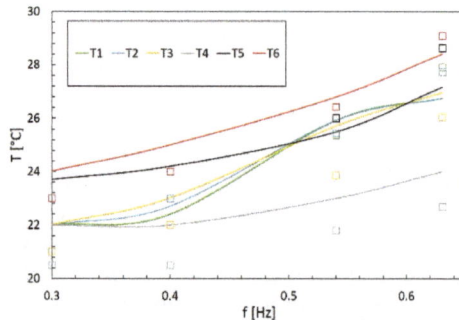

Figure 11. Surface temperature of the MCW in six different representative points as a function of operating frequency: experimental (symbols) vs. simulated (full lines) results.

4.2. Thermal Model Validation

To validate the thermal model, sub-valves outlet water temperatures (cold sub-valve ($T_{RV,C,out}$) and hot sub-valve ($T_{RV,H,out}$)) and their surface temperatures (cold sub-valve ($T_{RV,C}$) and hot sub-valve ($T_{RV,H}$)) have been measured and compared to model results for different operating conditions. An example of simulation results is reported in Figure 12 where the temperature distribution and heat flux are shown.

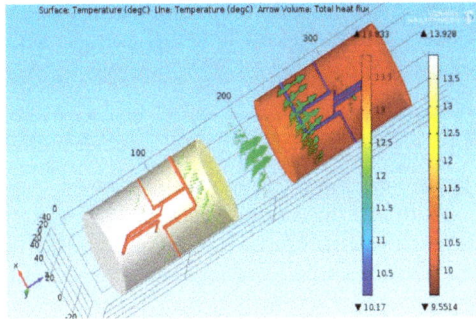

Figure 12. An example of simulation results for a rotating frequency of 17 Hz (that corresponds to 1000 rpm) and for a hot source temperature T_H of 22 °C.

In Figure 13, the comparison between the experimental and simulation results for the four temperatures mentioned above is reported. The value of the MAPE between simulation and experimental results is also shown for each temperature.

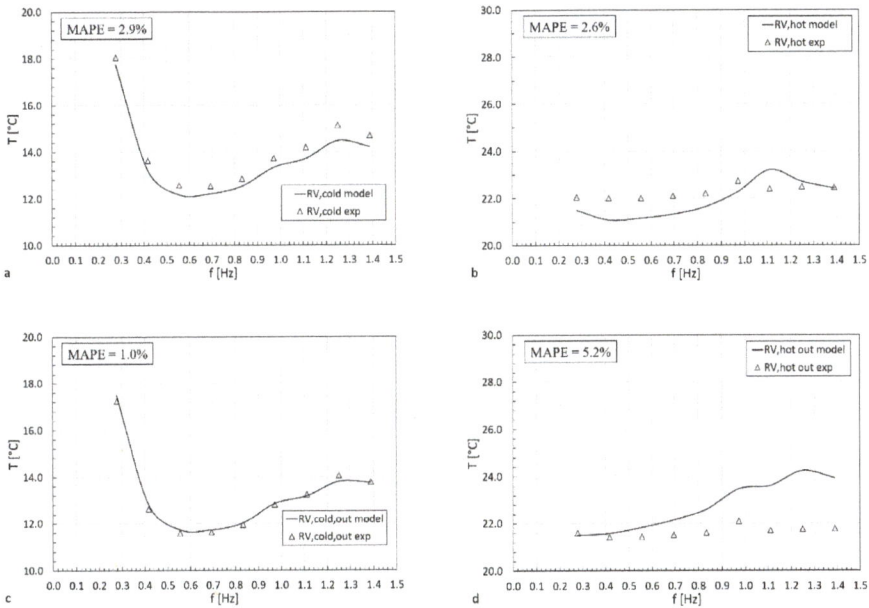

Figure 13. Comparison of experimental measurements and model results of (**a**,**b**) surface temperatures of both sub-valves ($T_{RV,C}$, $T_{RV,H}$) and (**c**,**d**) outlet water temperatures ($T_{RV,C,out}$, $T_{RV,H,out}$) for $T_H =$ 22 °C and different operating frequency.

The values of the relative error range and the maximum absolute error for each temperature are shown in Table 3. A maximum relative error of +11.4% occurs regarding $T_{RV,H,out}$. This large error could be related to the adiabatic surface hypothesis of the end parts of the rotary valve. However, these error values are considered acceptable considering the aim of this study.

Table 3. Absolute and relative error of the thermal model.

	$T_{RV,C}$	$T_{RV,H}$	$T_{RV,C,out}$	$T_{RV,H,out}$
Relative error range (%)	1.7–4.4	0.2–4.2	0.1–1.6	0.4–11.4
Absolute max error (°C)	0.7	0.9	0.3	2.5

In Figure 14, an example of result from the thermal model is shown. In detail, the parasitic thermal load $\dot{Q}_{c,loss}$ is evaluated as a function of the operating frequency (f) and hot source temperature (T_H). It is worth to notice that the parasitic thermal load presents a paraboloid shape with a maximum of about 61 W at an operating frequency of 0.8 Hz and T_H equal to 23 °C.

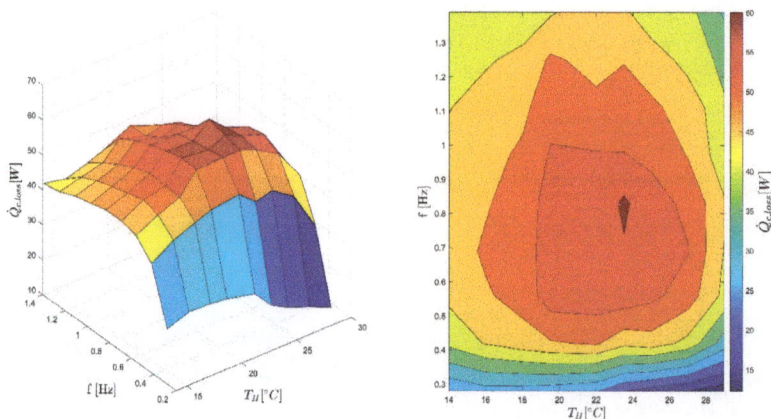

Figure 14. Parasitic thermal load of the rotary valve simulated by the thermal model with different operating frequencies and hot source temperatures.

5. Results and Discussion

The models developed in this work can help to quantify the main energy losses of '8MAG', and therefore to evaluate the performance which the prototype can achieve reducing at the minimum these losses. Before evaluating the achievable performance, a comprehensive energy characterization of '8MAG' was performed. In detail, cooling power, absorbed power and temperature span were measured for several operating conditions [15]. Then, it was possible to calculate the reference performance of the system, that is the coefficient of performance (COP), defined as

$$COP_{ref} = \frac{\dot{Q}_c}{\dot{W}_{tot} + \dot{W}_{pump}}. \tag{30}$$

The uncertainty of COP, measured by error propagation rules for indirect measurements, was estimated to ±0.28%.

Figure 15 shows the '8MAG' COP for different operating conditions, in terms of volumetric flow rate ($\dot{V}$), measured in L min^{-1}, and operating frequency (f) for three different levels of cooling power (50 W, 100 W, and 200 W) at a hot source temperature of 22 °C. From Figure 15, it is evident that COP decreases with the operating frequency increasing (at a constant cooling power) since the mechanical power required to move the magnets increases. The same trend can be noticed also considering the volumetric flow rate, even if it is less marked. However, the experimental tests showed that '8MAG' can achieve a maximum COP of about 2.5 with a cooling power equal to 200 W and a temperature span of 2 K. These results represent the baseline of the '8MAG' performance and they are used as a comparison to evaluate the possible improvements which could be achieved reducing the mechanical and thermal losses calculated by the models described in Section 3, regarding eddy currents and intrinsic thermal load of the rotary valve.

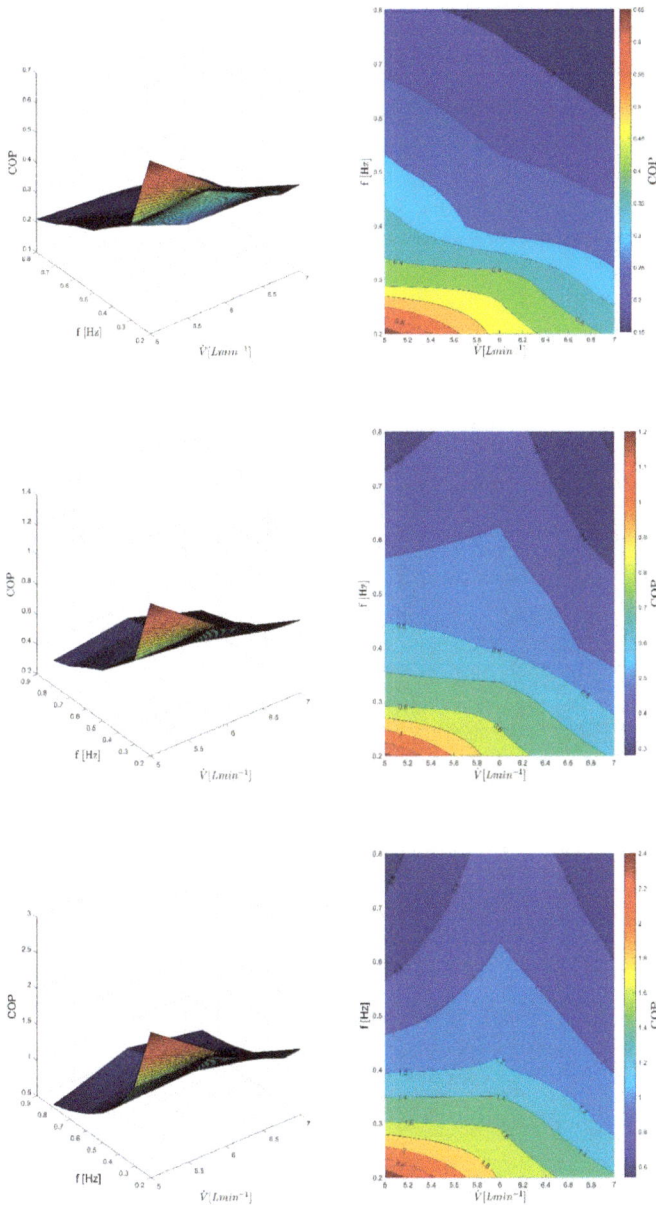

Figure 15. COP of '8MAG' as a function of volumetric flow rate and operating frequency at $T_H = 22\ °C$ and different cooling power (50 W, 100 W, and 200 W, from the top to the bottom). Temperature spans are in the range between 1 K and 8.5 K.

5.1. COP Improvement by Reducing Eddy Currents

The mechanical model allowed evaluating energy losses related to eddy currents generation in the MCW. These energy losses could be reduced by substituting the material of the MCW (or removing it). Hence, the new COP (COP_{ec}) can be calculated as

$$COP_{ec} = \frac{\dot{Q}_c}{\left(\dot{W}_{tot} - \dot{W}_{ec}\right) + \dot{W}_{pump}}, \tag{31}$$

where $\dot{W}_{ec}$ represents the mechanical power recovered by reducing eddy currents. In Figure 16, the COP improvement obtained with this solution is shown, both in absolute (ΔCOP) and relative (% ΔCOP) terms. It was calculated according to Equations (32) and (33).

$$\Delta COP = COP_{ec} - COP_{ref}, \tag{32}$$

$$\%\Delta COP = \frac{COP_{ec} - COP_{ref}}{COP_{ref}}. \tag{33}$$

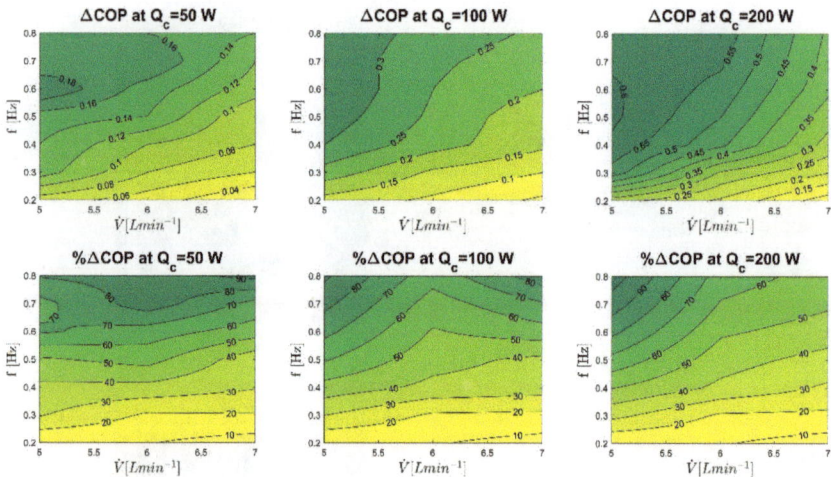

Figure 16. '8MAG' COP changing by reducing ECL on the MCW for different operating conditions at $T_H = 22\ ^\circ\text{C}$. Temperature spans are in the range between 1 K and 8.5 K.

It is evident that the COP improvement strongly depends on the cooling power ($\dot{Q}_c$), the operating frequency (f) and the volumetric flow rate ($\dot{V}$). As expected, the reduction of eddy currents leads to improve the '8MAG' performance for each operating condition. In detail, the improvement is more pronounced for higher cooling power, with a maximum COP increase of about 0.60, whereas it is about 0.18 with a cooling power of 50 W. Furthermore, it is worth to notice that the COP increase is more evident at higher operating frequencies, as one can see from the relative variation of COP (bottom plots in Figure 16).

5.2. COP Improvement by Reducing Parasitic Thermal Load

Using the thermal model, it was possible to evaluate the parasitic thermal load of the rotary valve, which is related to its non-adiabatic characteristic. Considering an ideal adiabatic rotary valve, and therefore trying to improve the insulation of this component, these thermal losses could be reduced and a new COP ($COP_{Q_{c,loss}}$) can be found as

$$COP_{Q_{c,loss}} = \frac{\dot{Q}_c + \dot{Q}_{c,loss}}{\dot{W}_{tot} + \dot{W}_{pump}}. \tag{34}$$

In Figure 17, the comparison between the actual COP of '8MAG' and the theoretical COP obtainable reducing the parasitic thermal load of the rotary valve is shown for different operating conditions. The data reported in Figure 17 are calculated according to Equations (32) and (33), substituting COP_{ec} with $COP_{Q_{c,loss}}$. It can be noticed that reducing the intrinsic thermal load of the rotary valve (improving its insulation) can improve the '8MAG' COP of about 0.40. It is worth to consider that the improvement does not change so much by varying the cooling power, in contrast with the improvement observed by reducing eddy currents losses. Furthermore, in this case, the beneficial effect increases with decreasing the operating frequency of the device.

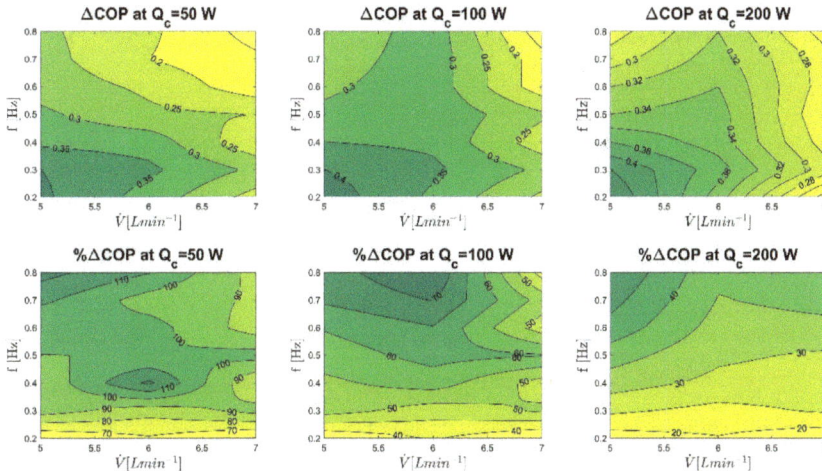

Figure 17. '8MAG' COP changing by reducing thermal losses in the rotary valve for different operating conditions at $T_H = 22\,°C$. Temperature spans are in the range between 1 K and 8.5 K.

5.3. Overall Achievable COP Improvement

To complete the analysis, the effect of introducing both the previous improvements (reduction of eddy currents generation in the MCW by substitution of the material and refinement of the rotary valve insulation) is investigated. Hence, another COP value ($COP_{ec+Q_{c,loss}}$) is calculated (see Equation (35)).

$$COP_{ec+Q_{c,loss}} = \frac{\dot{Q}_c + \dot{Q}_{c,loss}}{\left(\dot{W}_{tot} - \dot{W}_{ec}\right) + W_{pump}}. \tag{35}$$

As performed in the previous sub-sections, the comparison between this new COP and the reference COP of the prototype is shown in Figure 16, according to Equations (32) and (33).

Figure 18 shows the best obtainable COP of '8MAG' which can be achieved if both the energy losses investigated in this study are reduced. Hence, it represents the maximum improvements of the performance of '8MAG' for different operating conditions. In detail, a maximum increase of about 1.1 is highlighted with a cooling power of 200 W at relatively high frequencies (higher than 0.3 Hz) and low volumetric flow rate (lower than 6 L min^{-1}). Good improvements can be achieved also for lower cooling power, with maximum increases of the COP value of 0.65 and 0.8, with cooling power of 50 W and 100 W, respectively.

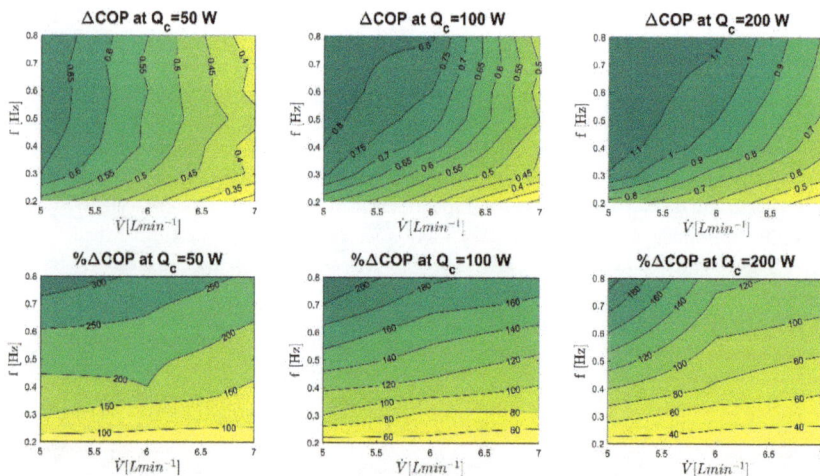

Figure 18. '8MAG' COP changing by reducing both thermal losses in the rotary valve and ECL on the MCW for different operating conditions at T_H = 22 °C. Temperature spans are in the range between 1 K and 8.5 K.

In Table 4, an overview of the results is shown, regarding the average, maximum, and minimum COP values in four different configurations, for three values of heat source temperature and cooling power. The average COP reported in Table 4 was calculated performing an average among all the COP values calculated at different operating frequencies and volumetric flow rates at the same heat source temperature and cooling power.

Table 4. Average '8MAG' COP for different cooling power and hot source temperatures (T_H). These values were obtained by an average among the experimental data at different volumetric flow rates and operating frequencies. Temperature spans are in the range between 1 K and 8.5 K.

	T_H [°C]	COP_{ref}			COP_{ec}			$COP_{Qc,loss}$			$COP_{ec+Qc,loss}$		
		Ave	Max	Min	Ave	Max	Min	Ave	Max	Min	Ave	Max	Min
$\dot{Q}_c$=50 W	16	0.35	0.68	0.15	0.45 28.6%	0.75 10.3%	0.29 93.3%	0.61 74.3%	1.08 58.8%	0.25 66.7%	0.79 125.7%	1.19 75.0%	0.46 206.7%
	22	0.35	0.68	0.15	0.45 28.6%	0.75 10.3%	0.29 93.3%	0.64 82.9%	1.09 60.3%	0.28 86.7%	0.83 137.1%	1.21 77.9%	0.51 240.0%
	32	0.38	0.69	0.15	0.49 28.9%	0.79 14.5%	0.28 86.7%	0.64 68.4%	1.04 50.7%	0.26 73.3%	0.85 123.7%	1.25 81.2%	0.48 220.0%
$\dot{Q}_c$=100 W	16	0.66	1.22	0.28	0.83 25.8%	1.35 10.7%	0.52 85.7%	0.94 42.4%	1.64 34.4%	0.37 32.1%	1.18 78.8%	1.82 49.2%	0.70 150.0%
	22	0.66	1.22	0.28	0.83 25.8%	1.35 10.7%	0.52 85.7%	0.97 47.0%	1.65 35.2%	0.40 42.9%	1.22 84.8%	1.83 50.0%	0.75 167.9%
	32	0.75	1.35	0.37	0.95 26.7%	1.51 11.9%	0.64 73.0%	1.02 36.0%	1.66 23.0%	0.56 51.4%	1.33 77.3%	1.87 38.5%	0.99 167.6%
$\dot{Q}_c$=200 W	16	1.52	2.53	0.89	1.81 19.1%	2.80 10.7%	1.21 36.0%	1.83 20.4%	2.96 17.0%	1.11 24.7%	2.18 43.4%	3.28 29.6%	1.51 69.7%
	22	1.52	2.53	0.89	1.81 19.1%	2.80 10.7%	1.21 36.0%	1.86 22.4%	2.97 17.4%	1.15 29.2%	2.23 46.7%	3.29 30.0%	1.57 76.4%
	32	1.58	1.97	1.21	1.83 15.8%	2.22 12.7%	1.44 19.0%	1.86 17.7%	2.26 14.7%	1.47 21.5%	2.16 36.7%	2.54 28.9%	1.76 45.5%

It is worth to highlight that the maximum '8MAG' COP, starting from the actual value of 2.5, can achieve a value equal to 3.3 (+30.3%), if the energy losses evaluated in this work can be reduced. However, only substituting the material of the MCW (to reduce eddy currents generation) can increase the COP up to 2.8 (+10.9%) whereas a better insulation of the rotary valve can increase the COP up to 3.0 (+17.5%). On the other hand, the second-law efficiency, starting from the previous maximum value of 2.4%, can achieve a value equal to 3.5% (+46%) at T_H = 16 °C and a temperature span of 4.7 K, if the energy losses evaluated in this work can be reduced.

6. Conclusions

In this study, the main energy losses of a rotary permanent magnet magnetic refrigerator, named '8MAG', developed at University of Salerno, were investigated with the aim to estimate the achievable performance of such a system. In detail, starting from the design details of the prototype, two different kinds of losses were identified: mechanical losses, regarding eddy currents generation inside the MCW and friction phenomena in the rotary valve, and parasitic thermal losses, related to the non-adiabatic conditions of the rotary valve.

Mechanical and thermal losses were investigated by developing two uncoupled models: the mechanical model, divided into three sub-models (SMF, SECP, and ST), which allowed to identify the power dissipated by eddy currents generation, and the thermal model, which could point out the parasitic heat load of the rotary valve. The mechanical power dissipated by friction phenomena was estimated by a semi-empirical model, based on experimental measurements and datasheets of the rotary valve components. The mechanical and thermal model were validated with experimental data, showing a good agreement with a maximum relative error of +8.0% and +11.4%, respectively. These models were developed for estimating only the main losses of the prototype to analyse the hypothetical achievable COP, and they do not consider the effect of temperature span, or even of the heat exchange between the MCW and regenerators, which could also affect the performance of the system.

Using the mechanical model, eddy currents losses were calculated for different operating conditions and maximum COP improvements were estimated within the range from 0.1 (with a cooling power of 50 W) to 0.3 (with a cooling power of 200 W). On the other hand, the reduction of parasitic heat losses, estimated by the thermal model, could lead to a maximum COP increase of about 0.5, showing a lower dependence on cooling power than eddy current losses. Reducing both eddy currents and parasitic heat losses, a greater improvement can be achieved, with a maximum COP increase of 0.8, which allows to reach a COP value of 3.3 (+32.0% against the reference value of 2.5). This increment allows '8MAG' to get closer to other magnetic refrigerator prototypes presented in the literature, in terms of performance. In detail, the improved '8MAG' performance are closer to the best performance showed so far by a magnetic refrigerator prototype (COP of 5 with a temperature span of 5 K), with a difference of 1.7 against 2.5 of '8MAG' without improvements.

The potential COP improvements showed in this study can be achieved on the real device by two actions: changing the aluminium of the MCW with another material with a lower electrical conductivity to reduce eddy currents generation and enhancing the insulation of the rotary valve to reduce parasitic heat losses. Future works could deal with experimental tests with the upgraded system, to test the actual performance improvement, and the analysis of other kinds of energy losses.

Author Contributions: A.M. (Angelo Maiorino) conceived the idea and helped with the discussion of the results and the model development. A.M. (Antongiulio Mauro) developed the model, performed experimental measurements, analysed experimental data and wrote some parts of the paper (when he was a Ph.D. candidate at University of Salerno). M.G.D.D. wrote some parts of the paper, prepared figures and helped with the discussion of the results. A.M.-B. helped with the discussion of results. C.A. supervised the entire work.

Funding: This research received no external funding.

Acknowledgments: Adrián Mota-Babiloni would like to thank the financial support from the Spanish Government through the postdoctoral contract Juan de la Cierva-Formación 2016 (ref. FJCI-2016-28324), and Banco Santander and Universitat Jaume I for the mobility grant "Becas Iberoamérica. Santander Universidades. Convocatoria 2018/2019".

Energies **2019**, *12*, 4388

Conflicts of Interest: The authors declare no conflict of interest.

References

1. Barclay, J.A. Theory of an Active Magnetic Regenerative Refrigerator. United States. Available online: https://www.osti.gov/servlets/purl/6224820 (accessed on 26 July 2019).
2. Steven Brown, J.; Domanski, P.A. Review of alternative cooling technologies. *Appl. Therm. Eng.* **2014**, *64*, 252–262. [CrossRef]
3. Rowe, A. Thermodynamics of active magnetic regenerators: Part I. *Cryogenics* **2012**, *52*, 111–118. [CrossRef]
4. Rowe, A. Thermodynamics of active magnetic regenerators: Part II. *Cryogenics* **2012**, *52*, 119–128. [CrossRef]
5. Aprea, C.; Greco, A.; Maiorino, A.; Masselli, C. The environmental impact of solid-state materials working in an active caloric refrigerator compared to a vapor compression cooler. *Int. J. Heat Technol.* **2018**, *36*, 1155–1162. [CrossRef]
6. Aprea, C.; Greco, A.; Maiorino, A.; Masselli, C. Magnetic refrigeration: An eco-friendly technology for the refrigeration at room temperature. *J. Phys. Conf. Ser.* **2015**, *655*, 012026. [CrossRef]
7. Greco, A.; Aprea, C.; Maiorino, A.; Masselli, C. A review of the state of the art of solid-state caloric cooling processes at room-temperature before 2019. *Int. J. Refrig.* **2019**, *106*, 66–88. [CrossRef]
8. Yu, B.; Liu, M.; Egolf, P.W.; Kitanovski, A. A review of magnetic refrigerator and heat pump prototypes built before the year 2010. *Int. J. Refrig.* **2010**, *33*, 1029–1060. [CrossRef]
9. Engelbrecht, K.; Pryds, N. Progress in magnetic refrigeration and future challenges. In Proceedings of the 6th IIF-IIR International Conference on Magnetic Refrigeration. International Institute of Refrigeration, Victoria, BC, Canada, 7–10 September2014.
10. Lozano, J.A.; Engelbrecht, K.; Bahl, C.R.H.; Nielsen, K.K.; Eriksen, D.; Olsen, U.L.; Barbosa, J.R.; Smith, A.; Prata, A.T.; Pryds, N. Performance analysis of a rotary active magnetic refrigerator. *Appl. Energy* **2013**, *111*, 669–680. [CrossRef]
11. Rowe, A. Configuration and performance analysis of magnetic refrigerators. *Int. J. Refrig.* **2011**, *34*, 168–177. [CrossRef]
12. Rosario, L.; Rahman, M.M. Analysis of a magnetic refrigerator. *Appl. Therm. Eng.* **2011**, *31*, 1082–1090. [CrossRef]
13. Romero Gómez, J.; Ferreiro Garcia, R.; Carbia Carril, J.; Romero Gómez, M. Experimental analysis of a reciprocating magnetic refrigeration prototype. *Int. J. Refrig.* **2013**, *36*, 1388–1398. [CrossRef]
14. Lozano, J.A.; Engelbrecht, K.; Bahl, C.R.H.; Nielsen, K.K.; Barbosa, J.R.; Prata, A.T.; Pryds, N. Experimental and numerical results of a high frequency rotating active magnetic refrigerator. *Int. J. Refrig.* **2014**, *37*, 92–98. [CrossRef]
15. Aprea, C.; Greco, A.; Maiorino, A.; Masselli, C. The energy performances of a rotary permanent magnet magnetic refrigerator. *Int. J. Refrig.* **2016**, *61*, 1–11. [CrossRef]
16. Aprea, C.; Greco, A.; Maiorino, A.; Mastrullo, R.; Tura, A. Initial experimental results from a rotary permanent magnet magnetic refrigerator. *Int. J. Refrig.* **2014**, *43*, 111–122. [CrossRef]
17. Lozano, J.A.; Capovilla, M.S.; Trevizoli, P.V.; Engelbrecht, K.; Bahl, C.R.H.; Barbosa, J.R. Development of a novel rotary magnetic refrigerator. *Int. J. Refrig.* **2016**, *68*, 187–197. [CrossRef]
18. Eriksen, D.; Engelbrecht, K.; Bahl, C.R.H.; Bjørk, R.; Nielsen, K.K.; Insinga, A.R.; Pryds, N. Design and experimental tests of a rotary active magnetic regenerator prototype. *Int. J. Refrig.* **2015**, *58*, 14–21. [CrossRef]
19. Huang, B.; Lai, J.W.; Zeng, D.C.; Zheng, Z.G.; Harrison, B.; Oort, A.; van Dijk, N.H.; Brück, E. Development of an experimental rotary magnetic refrigerator prototype. *Int. J. Refrig.* **2019**, *104*, 42–50. [CrossRef]
20. Albertini, F.; Bennati, C.; Bianchi, M.; Branchini, L.; Cugini, F.; De Pascale, A.; Fabbrici, S.; Melino, F.; Ottaviano, S.; Peretto, A.; et al. Preliminary Investigation on a Rotary Magnetocaloric Refrigerator Prototype. *Energy Procedia* **2017**, *142*, 1288–1293. [CrossRef]
21. Gimaev, R.; Spichkin, Y.; Kovalev, B.; Kamilov, K.; Zverev, V.; Tishin, A. Review on magnetic refrigeration devices based on HTSC materials. *Int. J. Refrig.* **2019**, *100*, 1–12. [CrossRef]
22. Plaznik, U.; Tušek, J.; Kitanovski, A.; Poredoš, A. Numerical and experimental analyses of different magnetic thermodynamic cycles with an active magnetic regenerator. *Appl. Therm. Eng.* **2013**, *59*, 52–59. [CrossRef]
23. Kitanovski, A.; Plaznik, U.; Tušek, J.; Poredoš, A. New thermodynamic cycles for magnetic refrigeration. *Int. J. Refrig.* **2014**, *37*, 28–35. [CrossRef]

24. Lucia, U. General approach to obtain the magnetic refrigeretion ideal coefficient of performance. *Phys. A Stat. Mech. Its Appl.* **2008**, *387*, 3477–3479. [CrossRef]

25. Trevizoli, P.V.; Nakashima, A.T.; Peixer, G.F.; Barbosa, J.R. Évaluation de la performance d'un régénérateur magnétique actif pour les applications de refroidissement—Partie I: Analyse expérimentale et performance thermodynamique. *Int. J. Refrig.* **2016**, *72*, 192–205. [CrossRef]

26. Aprea, C.; Greco, A.; Maiorino, A.; Masselli, C. A comparison between rare earth and transition metals working as magnetic materials in an AMR refrigerator in the room temperature range. *Appl. Therm. Eng.* **2015**, *91*, 767–777. [CrossRef]

27. Aprea, C.; Greco, A.; Maiorino, A. A dimensionless numerical analysis for the optimization of an active magnetic regenerative refrigerant cycle. *Int. J. Energy Res.* **2013**, *37*, 1475–1487. [CrossRef]

28. Gao, X.Q.; Shen, J.; He, X.N.; Tang, C.C.; Li, K.; Dai, W.; Li, Z.X.; Jia, J.C.; Gong, M.Q.; Wu, J.F. Improvements of a room-temperature magnetic refrigerator combined with Stirling cycle refrigeration effect. *Int. J. Refrig.* **2016**, *67*, 330–335. [CrossRef]

29. He, X.N.; Gong, M.Q.; Zhang, H.; Dai, W.; Shen, J.; Wu, J.F. Design and performance of a room-temperature hybrid magnetic refrigerator combined with Stirling gas refrigeration effect. *Int. J. Refrig.* **2013**, *36*, 1465–1471. [CrossRef]

30. Aprea, C.; Greco, A.; Maiorino, A. GeoThermag: A geothermal magnetic refrigerator. *Int. J. Refrig.* **2015**, *59*, 75–83. [CrossRef]

31. Lucas, C.; Koehler, J. Experimental investigation of the COP improvement of a refrigeration cycle by use of an ejector. *Int. J. Refrig.* **2012**, *35*, 1595–1603. [CrossRef]

32. Aprea, C.; Greco, A.; Maiorino, A. An application of the artificial neural network to optimise the energy performances of a magnetic refrigerator. *Int. J. Refrig.* **2017**, *82*, 238–251. [CrossRef]

33. Qian, S.; Yuan, L.; Yu, J.; Yan, G. Variable load control strategy for room-temperature magnetocaloric cooling applications. *Energy* **2018**, *153*, 763–775. [CrossRef]

34. Qian, S.; Yuan, L.; Yu, J. An online optimum control method for magnetic cooling systems under variable load operation. *Int. J. Refrig.* **2019**, *97*, 97–107. [CrossRef]

35. Aprea, C.; Cardillo, G.; Greco, A.; Maiorino, A.; Masselli, C. A rotary permanent magnet magnetic refrigerator based on AMR cycle. *Appl. Therm. Eng.* **2016**, *101*, 699–703. [CrossRef]

36. Lei, T.; Engelbrecht, K.; Nielsen, K.K.; Veje, C.T. Study of geometries of active magnetic regenerators for room temperature magnetocaloric refrigeration. *Appl. Therm. Eng.* **2017**, *111*, 1232–1243. [CrossRef]

37. Arnold, D.S.; Tura, A.; Ruebsaat-Trott, A.; Rowe, A. Design improvements of a permanent magnet active magnetic refrigerator. *Int. J. Refrig.* **2014**, *37*, 99–105. [CrossRef]

38. Monfared, B. Design and optimization of regenerators of a rotary magnetic refrigeration device using a detailed simulation model. *Int. J. Refrig.* **2018**, *88*, 260–274. [CrossRef]

39. Li, Z.; Shen, J.; Li, K.; Gao, X.; Guo, X.; Dai, W. Assessment of three different gadolinium-based regenerators in a rotary-type magnetic refrigerator. *Appl. Therm. Eng.* **2019**, *153*, 159–167. [CrossRef]

40. Klinar, K.; Tomc, U.; Jelenc, B.; Nosan, S.; Kitanovski, A. New frontiers in magnetic refrigeration with high oscillation energy-efficient electromagnets. *Appl. Energy* **2019**, *236*, 1062–1077. [CrossRef]

41. Czernuszewicz, A.; Kaleta, J.; Kołosowski, D.; Lewandowski, D. Experimental study of the effect of regenerator bed length on the performance of a magnetic cooling system. *Int. J. Refrig.* **2019**, *97*, 49–55. [CrossRef]

42. Capovilla, M.S.; Lozano, J.A.; Trevizoli, P.V.; Barbosa, J.R. Performance evaluation of a magnetic refrigeration system. *Sci. Technol. Built Environ.* **2016**, *22*, 534–543. [CrossRef]

MDPI

St. Alban-Anlage 66

4052 Basel

Switzerland

Tel. +41 61 683 77 34

Fax +41 61 302 89 18

www.mdpi.com

Energies Editorial Office

E-mail: energies@mdpi.com

www.mdpi.com/journal/energies

9 783039 219520